A Review of Research in Mathematical Education

Prepared for the Committee of Inquiry into the
Teaching of Mathematics in Schools

Part B   Research on the Social Context of Mathematics Education

A.J. Bishop and Marilyn Nickson

NFER-NELSON

Published by The NFER-NELSON Publishing Company Ltd.,
Darville House, 2 Oxford Road East,
Windsor, Berks. SL4 1DF.

First Published 1983
©A.J. Bishop and M.T. Nickson, 1983
ISBN 0-7005-0613-6
Code 8145 02 1

Printed in England

Distributed in the USA by Humanities Press Inc.,
Atlantic Highlands, New Jersey 07716 USA

Contents

Page

Chapter One The Institutional Aspect and   1
             Within-School Relationships
                 Relationships within primary schools
                 Leadership in primary schools
                 Relationships within secondary
                 schools
                 Leadership in secondary schools
                 The hidden curriculum
                 Subject status
                 Teaching resources
                 Summary

Chapter Two Pupils as a Constraint   15
                 Pupils' perceptions
                 Teachers' perceptions of pupils'
                 abilities
                 Pupils' language
                 The growth of pupils' attitudes
                 to mathematics
                 Summary

Chapter Three Societal Constraints   23
                 Parental expectations
                 Effects of parental involvement
                 The effects of class on pupil
                 achievement
                 Employers' views
                 Mathematics and the needs of
                 industry
                 Summary

Chapter Four The Structure of the Teaching   29
              Profession
                 The nature of teaching as a
                 profession
                 Primary teachers
                 Secondary teachers
                 Summary

Chapter Five The Effects of Initial Training of 37
              Teachers of Mathematics
                 Teachers in the primary sector
                 Teachers in the secondary sector
                 Summary

Chapter Six   Teacher Characteristics   43
                   Attitudes of mathematics teachers
                   Teachers' expectations of pupils
                   The effects of teachers'
                   perceptions of the mathematical
                   performance of girls
                   Summary
Chapter Seven In-Service Training and Professional   51
                   Development
                       In-service training
                       Professional developments
                       Summary
Chapter Eight Some General Conclusions   57
                   External constraints: the non-
                   autonomous teacher

                       Leadership roles
                       Support roles
                       The physical conditions
                       Summary

                   Internal constraints: the
                   individual teacher

                       Perceptions of content
                       Perceptions of pupils
                       Teaching individuals
                       Teacher stress

                   Outstanding problems
                   Concluding remarks
References   69
Appendix: Original Recommendations made to   79
               the Cockcroft Committee

Introduction

This book represents a part of 'A Review of Research in Mathematical Education' undertaken for the Cockcroft Committee of Inquiry into the Teaching of Mathematics in Schools. It reflects a sociological research basis in which neither mathematics, nor mathematics teaching is the essential focus. Rather, the concern is with the constraints (institutional and social) which surround the teaching of mathematics and with their effects upon teachers and pupils. As such, what we attempt to explore is the social context in which the teaching and learning of mathematics takes place.

In the course of this work, we were conscious of searching for sources of ideas rather than merely documenting research, and of relating them to the mathematical teaching and learning situation. Sometimes these sources were research reports and theses, sometimes D.E.S. or Schools Council surveys; sometimes they were actual submissions to the Committee. In order to help us understand and interpret this data, we also surveyed various writings and analyses, which were not necessarily based on any empirical research; neither were they always concerned specifically with mathematics teaching. However, in order to gain a clearer perspective, it is often valuable to step back from the immediate concerns of a problem. As will be seen, there exists a considerable body of relevant literature which is either directly or indirectly concerned with the issues and problems which surround mathematics teaching today. Our task has been to search out this literature, to integrate it into a meaningful whole, and to make recommendations based on our conclusions.

The analysis of this literature has been grouped into two main areas - that which concerns constraints external to the teacher and that which focuses more on what we have called internal constraints. Although there may be some overlap, we consider as external those constraints provided by the institutional organization, by pupils, by parents, by society and by the teaching profession. Internal constraints relate more to the teacher's own knowledge and attitudes, and to the roles of initial and in-service education. Consideration of the various external constraints appears in Chapters 1 to 4 inclusive, while Chapters 5, 6 and 7 examine the internal constraints. Throughout, we are concerned with relating our findings to both the primary and the secondary sectors in order to gain as clear a picture as possible at each level. In Chapter 8 we draw general conclusions from the evidence considered, in an attempt to identify the major issues which would appear to affect the social context of mathematical education in schools today. Finally, the recommendations which were included in our original report for the Cockcroft Committee appear in this book as an Appendix.

We are very grateful to many colleagues in Europe and America for their help and interest in the course of producing this work. We would also like to thank Maire Collins for her valuable contribution in typing the manuscript, and Sheila Hakin, librarian at the University of Cambridge Department of Education for her considerable assistance.

Alan J. Bishop
Marilyn Nickson

November 1982

Chapter One
The Institutional Aspect and Within-School Relationships

<u>RELATIONSHIPS WITHIN PRIMARY SCHOOLS</u>

Although few studies specifically concerning institutional effects on mathematics teaching have been carried out,
it is possible to infer from more general investigations, the
way in which such effects might act as a constraint in
primary schools. For example, points of relevance to
mathematics education emerge from a study by Ashton *et al.*
(1975) into aims in primary education in which 1513
practising teachers were involved. As well as being asked to
consider a list of 72 aims, teachers were asked to give a
relative weight to two descriptions of the fundamental
purpose of primary education. One of these was characterized as 'societal' (preparing the child to take his place in
society) and the other 'individual' (fostering the development of the child's individuality, interests and independence). Ashton related the societal description to a
traditional approach, and the individual description to a
progressive approach, in primary education. Teachers were
also asked to rate a range of five role descriptions on a
five point scale from 'strongly agree' to 'strongly
disagree'. The role descriptions given ranged from 'Most
Traditional' to 'Most Progressive' with 'Moderate' in the
middle. A further questionnaire to head teachers provided
information concerning school variables including the
school's environment and form of organization.

Possibly the most interesting result from the survey
was that each of the five role descriptions received a
response of 'agree' or 'strongly agree' from at least 40 per
cent of the sample, which suggests a considerable divergence
and spread in the way in which primary teachers view their
role. In relating teachers' opinions to school variables, it
was found that there was a 'lack of much strong relationship
between the characteristics of schools and the opinions of
the teachers working within them' (p.76). For example,
teachers in schools with a smaller staff only just preferred
a progressive teaching role more than those in a school with
a larger staff, and only a slight relationship was found
between vertical grouping and a preference for a more
progressive role. A stronger correlation could reasonably

have been expected between these two factors since small school size and vertical grouping have generally been associated with a progressive approach. It was also found that women tended to disagree strongly with traditional roles and preferred a moderate to a more progressive role significantly more than men. The general marked polarity of views between traditionalists and progressives was considered to suggest strongly that 'teachers' opinions about modes of teaching are firmly rooted in their fundamental views about the aims of education' (Ashton et al., 1975, p.55).

This conclusion is supported by Bennett (1976) who also found a strong relation between primary teachers' aims and teaching style. He, however, also noted some indication that primary school teachers feel that their traditional authority has been undermined by the reduction of classroom order that tends to accompany the adoption of more modern methods. It could be hypothesized that this feeling of loss of authority in a school with a progressive atmosphere may, to some extent, be responsible for the weakness of the relationship between teachers' aims and school variables quoted above. While teachers show a preference for the aims leading to the development of the individuality and independence of the child, they may at the same time find the pursuit of these aims difficult in a setting supposedly conducive to their achievement. Consideration however must be given to the fact that Bennett's (1976) work has received some criticism in terms of both design and terminology used in the study, to the degree where it has been questioned whether or not the results may validly be extrapolated to pupils in schools other than those in his sample (Wragg, 1976a; Gray and Satterley, 1976; McIntosh, 1979).

Whatever the reservations about this particular study may be, there does appear to be a link between the values reflected in teachers' aims and the way in which they teach. Value judgments play a crucial part in building the framework for teachers' decision-making and, clearly, much of teachers' decision-making is done in the context of lesson planning. As Clark and Yinger (1980) point out, 'As long as what a teacher is doing aids in *preparing a framework for guiding future action,* it counts as planning' (p.6, their italics). Insofar as planning is a manifestation of decision-making, it would seem that the judgments made by primary teachers in selecting what they teach would appear to show a degree of unawareness of the value aspect that determine those judgments, and hence how such values affect decisions. Values are determined by beliefs and, as Finlayson and Quirk (1979) note, ideology at the level of the individual is often referred to in terms of commitment to 'a belief in something' (p.52). The three areas of choice which polarized the views of the 1513 primary teachers in the Ashton et al. (1975) study as traditional and progressive were (a) the principles they employed in selecting curricular

2

content, (b) the way in which they involved pupils in learning, (c) the way they themselves promoted learning. Clearly teachers' aims must be related to beliefs as to what constitutes effective practice. If their beliefs can be seen to be manifested in the areas of choice referred to, then the value-laden nature of these three highly important aspects of their responsibility is clear. Where mathematics is concerned, without the identification of values as guidelines to rationalize their choices, teachers may take decisions which could produce extremes in terms of what mathematics is taught, and how. It could, for example, be a case of teaching only basic computational skills by mechanical, rote methods with little or no application to the problem-solving of everyday life, a tendency already noted by Ward (1979).

The problems faced by teachers in decision-making are discernible in the results of a recent survey, *Primary Education in England* (Great Britain, D.E.S., 1978a), in which Her Majesty's Inspectorate drew up lists of content items related to individual subjects which were 'likely to be found' in that area of study. These were items found to be considered by a substantial proportion of teachers as important, but as H.M.I. point out, 'They do not represent a full range of curriculum which is considered desirable or even necessarily a minimum curriculum' (p.77). They were selected on the basis of having appeared, individually, in at least 80 per cent of the classes surveyed. Where mathematics was concerned, only two-thirds of all classes in the survey were found to undertake work related to all of the items and when mathematics was grouped with English, less than two-fifths of the classes did all of the work identified in both subjects. 'This would seem to suggest that in individual schools either some difficulty is found in covering appropriately the range of work widely regarded by teachers as worthy of inclusion in the curriculum, or that individual schools or teachers are making markedly individual decisions about what is to be taught based on their own perceptions and choices or a combination of these' (p.80).

A similar situation would appear to exist in America. Freeman and Kuhs (1980), referring to such decisions faced by teachers of mathematics, state, 'Given that this teacher might also receive content messages through other sources such as district objectives, or comments made by parents, her principal, or other teachers, it is readily apparent that some of these messages must be ignored. Given restrictions in the time available for mathematics instruction, it is simply not possible to provide adequate coverage of all of the topics she will be asked to teach ... But what topics should she ignore?' (p.22). The dilemma posed by this question brings us to a consideration of leadership within primary schools.

Blyth (1965) notes that the norms selected by a head teacher 'may be affected by the educational tradition which is uppermost in his own attitudes' (p.98). For example, a head teacher might be characterized as 'progressive' in choosing to adopt a vertical method of grouping pupils according to age within the school or as 'traditional' in choosing to adopt a system of streaming. Whatever their tradition or attitudes, however, head teachers would reasonably be expected to exert a strong influence upon the teachers in their individual schools and hence, upon the curriculum (including mathematics). In doing so, they would be exercising their role as leader within the school. Morrison and McIntyre (1969) observe that head teachers are, indeed, 'sometimes referred to as leaders of the staff of their schools' (p.86). They argue, however, that two characteristics of a leader about which there is fairly general agreement are that (a) that person is a member of a group and, (b) they exert more influence upon the group than any other member. Accepting these characteristics as essential to the role of leadership, head teachers would be expected to have frequent contact with most members of their staff, otherwise they would effectively remove themselves from the membership of the group formed by the staff and without such contact, they would be unlikely to exert much influence. As Morrison and McIntyre (1979) put it, a head teacher who chooses to have little daily contact with most of his staff 'cannot be considered a member of the staff group or therefore its leader'.

In the study carried out by Ashton *et al.* (1975), head teachers of 201 primary schools were asked about the format of most of the consultations between themselves and their respective staffs. Approximately 69 per cent of schools in the sample had a full-time staff of five or more teachers and 30 per cent of the head teachers concerned were either full-time or nearly full-time in charge of a class. Of the 184 head teachers who replied to the question, only six had regular formal staff meetings and two had occasional formal meetings; 73.4 per cent reported frequent informal meetings and 21.7 per cent replied that they had both frequent informal staff meetings and occasional formal staff meetings. The net result, in the authors' words, was that 'even an occasional staff meeting was a feature of the organisation of only one-quarter of the sample schools.' (p.29)

Clearly, formal staff meetings in themselves do not constitute the only kind of contact that qualifies head teachers as members of the group formed by their staffs but there is a strong case to be made to support the contention that such meetings are necessary. The evidence gathered in the study indicates that a majority of head teachers (73.4 per cent) do not meet with their staff on a regular basis

4

but meet frequently and informally. It might be argued that frequent informal meetings may be sufficient to qualify them for group membership, but it is open to question how often head teachers might meet most members of staff on such an informal basis. Equally, it might be argued that approximately one-third of head teachers in the sample might qualify for group membership on the basis of their own teaching activities; however, it is more likely that their teaching may preclude contact with staff members because of other demands made on their non-teaching time.

With respect to the second criterion of leadership, it is doubtful whether frequent, informal contact would be adequate for head teachers to exert the degree of influence appropriate for the satisfaction of this criterion. Informal meetings imply irregularity. Without regularity, any contact could well lack the consistency that is desirable, if not necessary, for exerting the influence one would expect of a head teacher as a leader. Head teachers who actively engage with a class of their own would, without regular staff meetings, only be in a position to influence their staff by example since, as already noted, other duties would leave little time to be spent with staff to influence them in a manner which might best be described as of a professional development nature.

It is clear that the results of the study suggest that the leadership provided by head teachers of primary schools may vary to a considerable degree, especially when judged against the two criteria identified. Further evidence from the study indicates that the variation in the kind of leadership given can affect staffs in specific ways. It was found, for example, that where no formal meetings were held between the head teacher and the staff as a whole, 'teachers were significantly more likely to opt for a traditional role' (characterized as 'societal' insofar as it prepares the pupil for society) (Ashton *et al.*, p.79). Where regular staff meetings were held, 'teachers were more likely to choose more progressive roles' (characterized as 'individual' insofar as it was seen to foster the development of the pupil's individuality, interests and independence).

A related study, carried out in America in conjunction with a national evaluation of schools project, investigated the relationship between the type of 'administrative' leadership and pupil achievement in mathematics and teaching in elementary schools (Marcus *et al.*, 1976). Twenty-four schools were involved and data gathered through observation of classroom behaviour, interviews with school principals and self-administered questionnaires completed both by teachers and principals. Analysis of the data showed that in schools where the principals emphasized the importance of the selection of basic teaching materials and made more of the decisions with respect to the curriculum and teaching, there tended to be greater gains in pupils' achievement in both

5

mathematics and reading. These conclusions are further reinforced by Lezotte and Passalacqua (1978) who report studies which found that amongst the common features of schools characterized as especially effective was the fact that the principal had accepted responsibility for the instructional leadership of the school.

The other person, besides the head teacher, who could be in a position to exercise leadership in the mathematics teaching of the school is the teacher who holds the post of responsibility for mathematics, if it exists. However, although there has been a move to establish more of these posts in primary schools, the move apparently has not been entirely successful. The primary school survey found, for example, that in smaller schools (less than three-form entry) 'Posts with special responsibility for games were more common than posts for mathematics' (Great Britain, D.E.S., 1978, p.37). However, in schools where there were posts of responsibility, only in a quarter of these was there judged to have been a noticeable effect on the quality of the work throughout the school. This is borne out elsewhere where it is suggested by one Local Education Authority that while some of the people filling the posts of responsibility are effective, 'generally they do not exercise much influence over their colleagues or on the subject. In spite of the course held for them improvement in this respect is slow' (Cockcroft submission, B18) and the experience of another LEA indicates that 'there are few candidates able and prepared to take these responsibilities' (Cockcroft submission, B12). This raises the important question of teacher qualifications which will be dealt with in Chapter 6.

Her Majesty's Inspectorate (Great Britain, D.E.S., 1978a) do suggest that where positions of responsibility in mathematics were established, there was some evidence of these teachers 'planning programmes of work in consultation with the head, advising other teachers and helping to encourage a consistent approach' (p.37). It is to be hoped that with appropriate advisory support, this benefit can spread considerably wider than it apparently does at present.

RELATIONSHIPS WITH SECONDARY SCHOOLS

Turning to the secondary sector of education, a substantial amount of interest has been shown in organizational aspects of schools at secondary level since comprehensivization began some years ago. Studies such as those carried out by Richardson (1975), Rutter *et al.* (1979), Newbold (1977) and Francis (1975) all contribute to the building up of a general picture of how organizational characteristics combine to produce the atmosphere and ethos of the institution and, in turn, how they affect the teacher's role.

Rutter *et al.* (1979) were concerned with investigating school processes which they viewed as components of the social organization of the school and as creating the context in which teaching and learning take place. It may be helpful to identify each of these processes in order to appreciate the kinds of consideration they found to be important. They were (1) academic emphasis (2) teacher action in lessons (3) rewards and punishments (4) pupil conditions (5) children's responsibilities and participation in the school (6) stability of teaching and friendship groups and (7) staff organization. They found that the cumulative effect of these factors produced a particular ethos in a school and that the ethos differed from one school to another in terms of pupil achievement and behaviour. Apparently these differences were not related to physical aspects of the school nor to administrative considerations but rather to characteristics of the school as a social institution. This is of some interest in relation to mathematics education in view of expressed concern for the lack of physical facilities available for the teaching of the subject (e.g. Cockcroft submissions, B12, A41).

The social aspect of the school as an institution arises from the combined roles of the people in it, involving a complex of inter-personal relationships, pupil with staff, staff with staff, and pupil with pupil. Rutter *et al.* (1979) suggest that school processes are open to modification rather than being fixed external constraints, since they are controlled, to a greater or lesser degree, by various members of staff. For example, where teacher action in lessons is concerned, the teacher can supposedly decide what specific material is to be taught during a mathematics lesson. However, in reality, the head of the mathematics department will have decided the syllabus from which that material will be drawn and the head teacher may have decided that, in spite of the head of department's wishes, the class the teacher takes will be a mixed ability class. Each has been involved in decision-making at different levels but it is clear that it would be difficult for the class teacher alone to instigate change, and therefore the individual teacher is subordinate to the decisions of others. Thus while the constraints imposed may be open to change, they become increasingly rigid when they exist at the level of the class teacher and clearly have an effect upon how the teacher's role is carried out.

Hargreaves (1972), in his study *Interpersonal Relations and Education,* considers amongst other things, the teacher's role and how it may be implemented. He argues that, to be effective, teachers have to recognize the uniqueness of each and every teaching situation in which they find themselves and 'choose the role and style that (a) he can execute well, and (b) is the most appropriate to the pupils, the nature of the task and the general classroom situation' (p.153). Thus the learning situation is controlled by teachers in turn controlling their own role and it would seem to follow that the wider the range of role styles from which they have to draw, the better equipped they are to cope with the general classroom situation. Hargreaves highlights the importance of the teacher-pupil relationship when he suggests that what is required in taking any lesson, is a continuous diagnosis of the situation as the lesson proceeds so as to know when to shift roles. This, together with the ability to interpret accurately the feedback obtained, helps to gain some idea of the general effectiveness of the manoeuvre. For this to happen discipline must be maintained, hence instruction and discipline are based upon rules and norms that specify what is acceptable classroom conduct. They may either be imposed upon the pupils by the teacher or be agreed by teacher and pupils. For example, the mathematics teacher may impose a rule of strict silence and allow little movement within the classroom, or allow a degree of discussion and freedom of movement amongst pupils, beyond which they know they must not go.

Francis (1975) suggests that as the rules of a school in fact are drawn up by the head teacher, 'it is that distance from individual staff which makes them difficult to enforce' (p.148). In his book which, as he points out, is not based on empirical study but is the contribution of a practising teacher to the debate on discipline, Francis (1975) is concerned with presenting the teacher's picture of the classroom. He suggests that for pupils, the 'clearest expression of the school regime' lies in the rules, and that compulsory rules may sometimes present the individual teacher with an apparently insoluble dilemma (p.69). It may become a case of the teacher's survival being more important than the head teacher's approval, possibly the kind of situation envisaged when Hargreaves (1972) refers to rules negotiated between pupil and teacher. At such times an immediate solution must be produced which requires some accommodation on the part of the teacher. It must be stressed that this is one teacher's account of experience in one school, and it is possible to have considerable variation in the degree of staff involvement at rule-making level. However, this remains an example of how established rules and norms negotiated by others impose a constraint upon the teacher which may result in some crisis of conscience if ignored or altered.

LEADERSHIP IN SECONDARY SCHOOLS

The immediate arbiter of the institutional rules of the school for mathematics teachers is the head of department. The importance of the head of department's role is emphasized by one Local Education Authority which points out that 'one common element associated with quality is a good head of department with clear ideas and backed by a sound organisa-

tion' (Cockcroft submission, B25), a point further reinforced
by others (Neill, 1978; Cockcroft submission J61). Hall and
Thomas (1977), reporting the results of a study involving 39
mathematics departmental heads, describe their role as
'complex and obscure' and because of the ambiguous requirements, suggest that feelings of anxiety and job dissatisfaction as well as of futility and mistrust of colleagues tend
to build up. They appear to view their role not merely in
terms of academic demands but of managerial and representa-
tional demands as well.

The heads of department in this sample were happy to
accept that they represented the general ethos of the school
as determined by rules laid down by the head, but at the
same time, they chose not to hold regular meetings of their
department. This was interpreted as an indication that they
expected departmental members to accept their rulings, just
as they in turn had accepted those of the head of the school.
Since the managerial aspect of their role would seem to
include not only the organization of materials and personnel,
but also the direction of the department's aims and the
supervisory control of the work and standards of the department, it may be assumed that without regular departmental
meetings there could be little involvement of mathematics
staff in curricular decision-making of any sort.

There was also evidence that heads of department were
concerned to help unqualified members of staff, but 'they
were neither enthusiastic about the value of formal
departmental meetings for this purpose nor prepared to accept
automatically responsibility for the discipline problems
faced by a probationer teacher' (p.35). Although these
results are based on a small sample, they do provide some
indication of the complexities and problems entailed in the
head of department's role. A further study carried out by
Hall and Thomas (1978) into the role specification for heads
of mathematics departments as sent to applicants for such
posts by schools, suggests that the complexities of the role
have yet to be understood or identified by head teachers
themselves.

An indication of the potential influence of the head
of department's role is given by Hargreaves (1967) in
connection with factors affecting the attitudes of teachers.
This was that the system of allocation of teachers to
particular classes tended to be taken by teachers as an
indication of their basic competence or incompetence. The
teacher given a C stream mathematics class (or a lower set)
to teach might consider this a manifestation of the head of
department's judgment about his or her general competence as
a teacher. There may no doubt be situations in which ability
groups are spread as evenly as possible over members of a
department and teachers may even be consulted about the
choice of groups they will teach. However, it must be

recognized that there may also be instances where this does
not occur.

The hidden curriculum

The institutional aspects of the school as represented
to mathematics teachers by their heads of department form
part of what has become known as the hidden curriculum. The
idea of a hidden curriculum is one which is open to wide
interpretation and which, at times, may be in danger of
becoming confused simply because it is so diffuse and
encompasses so much. At its simplest, Gordon (1978) sees
the 'distinction between the explicit and the "hidden
curriculum"' as being those factors which link what is taught
(the explicit curriculum) with the organization of the
school (p.248). He suggests that it relates to 'the basis
of organising pupils - whether it be streaming, banding or
mixed ability - and the structure and legitimation of hierarchies in schools' (p.248). In Apple's (1980) words, 'We
see schools as a mirror of society, especially in the
school's hidden curriculum' (p.1). In terms of the study
carried out by Hargreaves in 1967, the hidden curriculum
would relate to the fact that the school was a selective
school for boys, that it was streamed and had a non-
academic, custodial atmosphere, that there were few extra-
curricular activities and that there was culture clash
between staff and pupils. Clearly the structure and
hierarchies that are implicit in organization factors of
this nature would enter, to a greater or lesser degree, into
the teaching and learning situation concerned with the
explicit curriculum. The hidden curriculum may pervade the
classroom through what may be accepted, unquestioningly, as
ordinary organizational procedures such as the grouping of
pupils for purposes of teaching or whether or not there is a
school uniform, while they are in reality procedures
selected at a higher level to promote the ethos of the
school. An example of this might be where, in a school that
places highest priority on academic excellence, pupils may
be rigidly streamed and the head of the mathematics
department will teach the top stream classes only.

The hidden curriculum is at present coming under deeper
scrutiny and analysis. Apple (1980) questions whether
schools *are* merely 'reproductive mirrors' and suggests that
if, as in other work areas, there are in schools 'elements
of contradiction, of resistance, of relative autonomy' then
they have 'transformative potential' (p.22). It is possible
to extrapolate such a view to the level of the mathematics
department or even the individual mathematics classroom and
imagine the existence of such elements and postulate their
effects. Thus at departmental level, members of a department
could conceivably resist a head of department's determination
not to involve them in curricular discussion and policy by
becoming involved with each other as a group, to engage in

such discussion and self-help. At the level of the mathematics classroom, both teacher and pupils could, to a greater or lesser degree, tend to resist the apparently 'given' aspect of the hidden curriculum and the control which it exerts. Indeed, it could well be to the advantage of the teaching and learning of mathematics if some of the unquestioned assumptions with regard to what constitutes a 'good' mathematics classroom were to be challenged in just such a way.

Subject Status

A teacher's specialist subject also has a part to play in the value judgments made by other teachers, and mathematics is particularly notable in this connection. In an investigation into *Authority and Organisation in the Secondary School* carried out for the Schools Council, Richardson (1975) pinpoints mathematics as a subject compartmentalized as 'academic' and suggests that if the situation were otherwise, it could lead to the possible release of 'unexpected talents in children and corresponding skills in teachers' viewed in terms of a kind of creative potential (p.41).

This conclusion is one that might well be disputed by mathematics teachers themselves who may possibly derive a certain amount of enjoyment, not to mention kudos, from the 'academic' nature of their subject. On the other hand, those mathematicians who support strongly the aesthetic and creative aims of education might more readily agree with Richardson. Musgrave (1979) reminds us, with respect to the curriculum, that the content of the whole collection displays values and Gordon (1978) takes this further when he suggests that bringing together this collection 'raises questions relating to the status of subjects: whether it is "given" and if there is any logical distinction between high status (mathematics and science) and low status (social studies and economics) subjects' (p.148). Morrison and McIntyre (1969) point out that a potential source of conflict among teachers within a school is the status perceived by one teacher or department to be offered another teacher or department.

Traditionally it would seem that mathematics has been attributed high status as a discipline and mathematicians, accordingly, have been assumed to enjoy status equal to their subject. This is reinforced to some extent by the fact that mathematics, arguably, is the only subject that has not been integrated with another in the curriculum (even though 'used' in other subjects, it is normally not taught in integration with them). Thus mathematicians have been able to retain an individual identity as 'an authority' unlike some historians, for example, who may have lost theirs in the thicket called 'humanities'. One can thereby see how the status ascribed to

mathematics teachers by their subject may become a source of
friction with other colleagues. This identity has possibly
a more marked effect with respect to the pupils' perception
of the mathematics teacher, as we shall come to see.

Apart from the question of subject status, a further
potential source of conflict with fellow colleagues is the
fact that other subject specialists sometimes claim that the
demands made upon mathematics by their subject are not always
met. This is discussed in relation to physics by Belsom and
Elton (1974) and to biology by Dudley (1975).

Teaching resources

Finally, further constraints that arise from within the
school are related to the resources and accommodation
provided for the teaching of mathematics. Although mixed-
ability teaching has become more prevalent and with it, the
increased demand for a wider variety of resources (Lingard,
1976), these apparently are not always forthcoming. It
appears to be something of an anomaly that mathematics, with
its supposedly high status within the curriculum, should be
one of the few subjects that apparently very often does not
have a specialist teaching area. For example, one Local
Education Authority reports only two out of fourteen
secondary schools as having special mathematics centres
(Cockcroft submission, B12) while another has less than 3
per cent of secondary schools that have a room that is
mathematically equipped (Cockcroft submission, B31). It has
been suggested that this 'nomadic existence' of the teacher
moving from one area to another has resulted in something of
an ad hoc approach to the teaching of the subject, giving
rise to little concern for display of pupils' work and for
the use of good materials (Cockcroft submission, J52).

Display of the mathematical work of pupils in secondary
schools was found to be lacking in the recent D.E.S. (1979b)
survey, with only 40 per cent of schools considered to have
made any effort at all. There was no specially equipped
room for mathematics in 66 per cent of schools surveyed,
while statistical analysis revealed that where there was
specialist accommodation, then display, practical work and
the use of games and puzzles were more likely to be found.
An important feature noted with respect to grouping rooms
together was that the head of department was enabled
generally to support, and to supervise, the work of his
colleagues. While it was acknowledged that specialist rooms
suitably grouped clearly increased the quality of the
teaching of mathematics, 'the allocation of such rooms is no
guarantee that the opportunities they offer will be taken up'
(Great Britain, D.E.S., 1980a, p.14). Why this may be so, one
can only surmise. However, it would seem that given favourable circumstances of this nature, the job of the head of
department in setting an example and leading in the effective

use of resources could be made easier.

It is interesting to note that the survey also showed that in only 23 per cent of all schools did pupils in years 4 and 5 have experience with computers. Reports from America suggest that a secondary school without a computer terminal would be most exceptional.

SUMMARY

Consideration of institutional features of the primary school indicate that the head teacher has major control over matters of organization (Blyth, 1965). Evidence suggests that (a) primary head teachers have little formal contacts with their staffs, and (b) where there is little formal contact of this kind, teachers tend to adopt a traditional approach in the classroom (Ashton *et al.*, 1975). It is suggested that this may, in part, reflect a type of leadership on the part of head teachers. Posts of responsibility for mathematics in primary schools would appear not to have made much impact on the quality of mathematics teaching as yet.

The institutional factors of the secondary school that affect the curriculum, viewed from the perspective of school processes, are seen to be open to change (Rutter *et al.*, 1979). However it would appear to be increasingly difficult for this to happen at the level of the individual class teacher since the rules of a school are drawn up by the head (Francis, 1975) and are mediated through the head of department (Hall and Thomas, 1977). There are indications that heads of mathematics departments do not involve mathematics staff in matters of curricular decision-making and that they tend to view their role as being ill-defined and hence with some dissatisfaction (Hall and Thomas, 1977). The effectiveness of mathematics departments has been found to be directly related to the leadership given by heads of department and an increase in formal departmental meetings in schools is advocated (Great Britain, D.E.S., 1979b). The power of the head of the department is sometimes assumed be teachers to be manifested in their allocation to particular classes which they see to be a reflection of the judgment of their competence by the head of department (Hargreaves, 1967).

The highly structured organization of the secondary school gives rise to a complex and potentially powerful hidden curriculum. Where mathematics is concerned, this hidden curriculum may manifest itself through superior status ascribed to the subject which could cause resentment amongst members in other departments (Morrison and McIntyre, 1969).

However, despite this supposedly superior status, it seems to be the case that fewer than one-third of all secondary schools have minimal resources for adequately teaching mathematics (Great Britain, D.E.S., 1980a).

Chapter Two
Pupils as a Constraint

No doubt the most important external constraint upon teachers
is the pupils they teach. The effect of pupils must be
identified not only in terms of what they, as individuals,
bring to the mathematical learning situation, in terms of
intellectual development, capacity for learning and past
mathematical experience, but also in the light of the part
they play in the social 'arena' of the classroom. Clearly,
while most of these effects are common to both the primary
and secondary level, some may be exerted more strongly at
one level than at the other. In either case, however, as
Nash (1973) suggests, 'All genetic and sociological factors
are mediated and realized through the interaction between the
teacher and the child in the classroom' (p.123).

PUPILS' PERCEPTIONS

   The mutual perception of teacher and pupil is a paramount factor in the interactive situation within the
classroom. These teacher-pupil perceptions are highly
complex at primary level, as not only does the primary school
child rapidly progress through a variety of stages in social
development, but also the primary school teacher's role
achievement, of necessity, depends upon adaptation to those
stages (Blyth, 1965). To use Blyth's example, there is a
difference between the seven-year-old's perception of the
teacher as an authority figure and the nine-year-old's, and
a resultant difference in their reaction to any sign of
weakness on the teacher's part. With the younger children,
the reaction would be one of 'bewildered anarchy' while
the older ones would present the teacher with a kind of
corporate hostility. The younger pupils' expectations are
that order will be maintained while the older pupils expect,
amongst other things, efficiency; in short 'they want to
have a fitting object for their loyalty and identification'
(p.102). Added to feelings of this kind, is the anxiety
level identified by Trown and Leith (1975) as a distin-
guishing factor between those who do and do not benefit
from a learner-centred approach. They found a teacher-
centred, supportive strategy, on the other hand, to be almost
equally effective whatever the level of anxiety. Bennett
(1976) also found that in the teaching of mathematics, more

15

than any other subject, teaching style appears to have a stronger effect on pupil achievement and that gains seem to be greatest where a formal, teacher-centred style is used. Thus at primary level the establishment of norms takes place against a background of continually changing social and emotional relations, as well as the demands of psychological nature with respect to the individual learning styles of pupils.

In a study of a class of twelve-year-olds during their first term in secondary school, Nash (1973) attempted to identify how pupils tend to discriminate between different teacher behaviours. He found that six pairs of constructs emerged strongly in the way in which pupils described how teachers behave. These were: (1) Keeps order - Unable to keep order, (2) Teaches you - Doesn't teach you, (3) Explains - Doesn't explain, (4) Interesting - Boring, (5) Fair - Unfair, (6) Friendly - Unfriendly. He suggests that the identification of these constructs shows how clearly 'the pupils' view of what is appropriate teacher behaviour and what is not is well developed' (p.50). The interesting observation is made that the pupils' conception of their own role is a passive one, in which they do not see themselves as actively finding things out for themselves or attempting to control their own behaviour. Nash (1973) concludes, 'If the experience of school does generate such limiting self-definitions it is surely not wholly achieving its aims' (p.58).

Hargreaves (1972), in examining interpersonal relationships at secondary level, notes that pupils tend to share a generalized attitude towards the teacher and he classifies teachers as direct or indirect according to the degree to which interaction between pupil and teacher takes place. Here he draws on the work by Flanders (1973) where, in studying relations among teachers, pupils and their attitudes he classified teachers as direct and indirect, according to the preponderance of the kind of statements made by teachers to pupils. The direct teacher tends to be a purveyor of information while the indirect teacher is seen as pupil-centred, allowing the initiation of ideas to come from the pupils. It is suggested that the indirect teachers produce better attitudes to learning and higher attainment on the part of pupils; the teacher who takes into account the ideas and feelings of pupils is rated as 'good'. The study carried out by Yates (1978) of four mathematical classrooms provides examples of what could be identified as 'indirect' and 'direct' teaching and the reactions of pupils to the different approaches. One teacher quoted (who could be characterized as 'indirect') uses an open question and pupils' subsequent answers and further questions to develop the idea of the process of elimination in linear programming. She notes that 'He is not afraid to listen to the pupils' interpretation of questions' (p.115). On the other hand, the dialogue between another teacher and his pupils indicates that 'he expects the pupils to get on to his line' by posing questions that have a highly specific answer, and by interrupting what are obviously wrong explanations offered by pupils so that 'the pupils sit there tolerating him, endeavouring to find his answers at appropriate moments' (p.107). Hence the pupils react according to the value they perceive the teacher to place on their contribution to discussion.

Another idea concerning pupil expectations of teachers is contained in Skemp's (1979) discussion of the different goal structures which pupils and teachers may hold. These goal structures are seen mainly in terms of two kinds of understanding. Firstly, 'Instrumental understanding, in a mathematical situation, consists of recognising a task as one of a particular class for which one already knows a rule' (p.259). The second, relational understanding, on the other hand, is seen to consist mainly of relating a task to a suitable schema. In instrumental understanding or learning, the goal is simply for the pupil to get the right answer, while in relational learning, the goal is more complex and the teacher seeks some indication that the pupil can fit what has been learned into an appropriate schema, thus indicating that they not only know what is right, but why it is right as well. Clearly, in one such mismatch the pupil may be concerned only with obtaining the correct answer while the teacher is going beyond this and trying to work towards establishing a schema. The pupil's reaction might well be to 'switch off', thus causing an adverse effect on his attitude. The mismatch may also be reversed; the pupil may search for reasons why something is the case, and may therefore attempt to develop relational understanding, while the teacher, perhaps with an insecure knowledge of mathematics, will ignore the questions and persist at the instrumental level. Again, the pupil's attitude will deteriorate. Both of these descriptions of conflicting pupil-teacher expectations indicate how the development of genuine interest on the part of pupils may be obstructed and how the lack of meaningful dialogue between them and their teacher may result. Perhaps this is the type of situation which gives rise to the following kind of statement: 'For the majority of schools, mathematics is a rather dull routine business both for teachers and children' (Cockcroft submission, B18).

TEACHERS' PERCEPTIONS OF PUPILS' ABILITIES

A further constraint exerted by pupils upon teachers at both primary and secondary level concerns their different abilities. Teachers apparently find such differences not only recognizable but also significant for their teaching, as in approximately three-quarters of the primary schools surveyed by Her Majesty's Inspectorate, children were grouped for ability in mathematics within classes at the ages of 7, and 11 years (Great Britain, D.E.S., 1978a). This teacher response to the presence of different pupil abilities has certain effects, of course, with one judgment being made in the survey that the more able were not being adequately extended. Perhaps more concern and time is being given by the teachers to the average and less able pupils at the expense of the more able who could be considered to be more 'self-sustaining'. Another prerequisite for the successful stretching of more able pupils is the teachers' confidence in their own mathematical knowledge, and this factor will be discussed more in Chapter 6.

The situation in secondary schools is only marginally different. In the secondary survey carried out by H.M.I. (Great Britain, D.E.S., 1979b) half the comprehensive schools had some form of ability grouping in the first year, but the figure had risen to over 90 per cent by the third year.

We must remember, however, that what teachers are responding to is essentially their *perceptions* of the pupils' abilities. Hargreaves (1967) sensitized us to this point when he studied how both teachers' and pupils' attitudes and behaviour developed in the course of their adaptation to the system of a streamed, secondary modern school over a period of four years. He argues that because of the minimal contact most secondary teachers have with their pupils, the teachers' assessments of them tend to be more indirect and based upon their expectations of role-conformity on the part of pupils, as opposed to being based upon frequent and more direct personal contact with them. From this there follows a cat-egorization of pupils by teachers on minimal evidence and any future teacher-pupil interaction will be defined by this categorization. Thus begins the self-fulfilling prophecy where, as Hargreaves (1967) suggests, the pupil will adjust to the teacher's categorization by exhibiting behaviour appropriate to it. Hence, pupils in mathematics classes who may in fact have reasonable mathematical ability but who may give incorrect answers orally in class from sheer nervousness, may too easily be labelled as incompetent and, as a result, may give up any effort to develop what mathematical ability they have.

Further evidence about teachers' perceptions and expectations comes from Nash (1973). He followed his sample of pupils from primary school through to secondary school, and found that in evaluating pupils, teachers used personal constructs rather than academic ones, the three most common being Hardworking - Lazy, Mature - Immature and Well behaved - Poorly behaved. This may be interpreted as being due to the lack of any great amount of direct teacher-pupil contact already identified by Hargreaves. If pupils do, indeed, conform to the kind of categorization that labels them 'poorly behaved', then the self-fulfilling prophecy becomes a vicious circle and the weak teacher is likely to be faced with discipline problems. One consequence of this which Nash (1973) suggests, is that the first thing pupils expect of teachers is an ability to keep order and if they are not capable of doing so, they are regarded by the pupils 'as having broken the rules' (Nash, 1973, p.128). Thus, he argues, the intransigent pupil will feel justified in behaving disruptively.

This type of problem may be of particular interest in relation to mathematics teachers since there is some evidence that more mathematics teachers have their probationary year extended than is the average in other subject areas (Cockcroft submission, B18). This feeling is also expressed in further evidence which states that head teachers are 'concerned about the poor quality of mathematics probationers, far too may of them passing only marginally at the end of their first teaching year' (Cockcroft submission, J61). Although Francis (1975) suggests that to claim that class-room control is strongly allied with the subject being taught is to claim 'dubious foundation' for it, he does acknowledge that the subject is important (p.70). In the case of mathematics, where pupils may already have been labelled 'badly behaved', the visibility of success or failure which is inherent in 'doing' mathematics may clearly exacerbate an already difficult situation. This may work both ways, of course, as exemplified by a pupil identified by Hargreaves (1967) as saying 'Mr X is the best of all teachers 'cos he makes maths so simple and easy' (p.94). However, there is also the worrying fact that a substantial amount of mathematics appears to be taught by non-mathematicians. For example, one authority indicates that just over half of their mathematics staff teach the subject full-time and only 72 per cent of those have 'approximately suitable training in the subject' (Cockcroft submission, J52). Thus the situation arises where teachers with a poor grasp of mathe-matics, who teach it with little authority, are also likely to be faced with a loss of authority in the disciplinary sense.

PUPILS' LANGUAGE

Pupils' language and the extent to which the code they use is restricted or elaborated has been recognized as an important factor in classroom learning generally (Bernstein, 1971). Barnes (1971), however, reminds us that it is not known to what depth personality patterns have already been determined by the time the child first enters school and the extent to which they may be changed by new language experience in order to overcome any 'restrictive' character-istics. In any case, the pupil's language has been recog-nized as of particular importance for mathematical learning. At primary level, the difficulties arising are compounded by the fact that the children, faced with written mathematical schemes of learning, are at the same time in the throes of learning to read. Shuard (1979) has drawn attention to the kinds of problem raised in this connection, including such matters as styles of writing, visual material used and the ease with which ambiguities arise.

The recent D.E.S. primary survey (Great Britain, D.E.S., 1978a) points to an even greater problem in that a third of all the schools in their sample had some children for whom English was the second language. With respect to English as the weaker language in mathematical learning, Dawe (1978) reviews the general conclusions from research that establish points such as the pupil's inability to group word meanings quickly, and the crucial cumulative effect of the many factors that go into the teacher's handling of the teaching-learning situation, and how they bring them to bear on coping with the bilingual child. He suggests also that most studies in this field have tended to dwell on the effects of bilingualism on the pupil's mathematical performances in mechanical arithmetic, while few have attempted to study its effect on thinking processes which underlie the learning of mathematics. Clearly, whatever the constraints of the child's language on the teacher's effectiveness, this is heightened considerably in the case of the bilingual child.

THE GROWTH OF PUPILS'S ATTITUDES TO MATHEMATICS

As pupils develop throughout the different phases of schooling they become increasingly aware of mathematics as a subject and this awareness clearly affects the growth of their attitudes to mathematics. The primary/secondary transition is a point where this becomes evident. Newbold (1977) for example, cites the difficulties arising from the variation of their primary experience that pupils coming from different primary schools bring to their common secondary school. Mathematics is mentioned in particular where marked differences in the performances of pupils noted at the end of the first year were related to the primary schools from which they had come. This is reflected, to some extent, in the results of the survey of primary schools referred to earlier, where it was noted that only two-thirds of all classes in the survey included work related to all of the items identified by teachers as forming a mathematics curriculum. As a result of the kind of mathematical experience they will have had at primary level and, more particularly, their achievement or lack of it with respect to the subject, attitudes to it are likely to be entrenched by the time they enter secondary school.

Evidence suggests that at lower secondary level few pupils like mathematics, but they do recognize its useful-ness and the necessity of having at least some knowledge of it. Duckworth and Entwistle (1974) found also in investigating the attitudes of 600 second-year and fifth-year grammar school pupils that of nine subjects studied, mathematics was rated seventh for interest by the fifth year, and fourth for difficulty. Even when studied at Advanced Level, adverse attitudes persist and continue to deteriorate and, somewhat surprisingly where favourable attitudes exist, these are not mirrored in higher achievement (Selkirk, 1974).

Another point of transition at secondary level is when pupils choose, for the first time, the subjects they will pursue to a higher level. Musgrave (1979) considers this aspect of choice to be one of the more important features of secondary schooling, and sees it as being complicated by an increased development in pupils' self-awareness. He quotes Hudson (1968), who postulates that, at this age, pupils are able to differentiate among four different selves and, as a result, reflective choice becomes difficult for them. The four selves are the 'actual self', which is who they really are, the 'ideal self', which is who they would like to be, the 'perceived self', the person their teachers perceive them to be, and the 'future self', the person they expect to be in a few years hence (Musgrave, 1979, p.229). All four are seen to be interdependent and to affect the ways in which pupils make their choices. This point is illustrated by suggesting that the choice made by pupils in selecting the subjects they will study determines the path their future will take but, at the same time, it affects their teachers' perception of them. This is reinforced by Selkirk (1974) in his study of pupils' choice of mathematics as an Advanced level subject. He interpreted a surprisingly lower ranking of dislike for mathematics as a subject in one particular survey, compared with results of other similar surveys, as being due to the fact that the head of department was administering the questionnaire. He surmised that the pupils were concerned about their future relationships with the staff of the mathematics department if they were to indicate a dislike for the subject. (This was in spite of the guaranteed confidentiality of the results.) Thus the pupil's perception of the teacher remains a major influence and possibly, in some ways, becomes a more subtle constraint at secondary level.

SUMMARY

From a social perspective, the effectiveness of mathematics teachers appears to be constrained predominantly by pupils' expectations, abilities, attitudes and language. It would appear that pupils' perceptions and expectations of teachers are well defined both at primary and at secondary level (Blyth, 1965; Hargreaves, 1972; Nash, 1974). As well as developing more general expectations, pupils in mathematics classes may well be seeking different goals to those being pursued by the mathematics teacher (Skemp, 1979).

Concerning pupils' abilities, whatever these may be in reality, the teacher's perception of pupils' mathematical ability is of paramount importance (Hargreaves, 1967). These perceptions ultimately determine the grouping of children (Great Britain, D.E.S., 1978a; 1979b), the type of teaching they receive and more importantly, the pupil's self-picture, particularly with respect to their mathematical abilities. Pupil's language is also recognized as an important factor in the learning of mathematics, particularly in cases where English is the pupil's second language (Great Britain, D.E.S., 1978a; Dawe, 1978).

As pupils progress through schooling, their perceptions of mathematics as a subject become crystallized. Different mathematical experiences at primary level result in a variation in attitude and achievement at secondary level (Newbold, 1977), which critically affect the pupil's choice of subject for further study.

22

Chapter Three
Societal Constraints

Having considered in the two previous chapters the constraints on the teacher from within the school, we now look
outside the school to the pressures which come from the
wider society. Regarding mathematics teaching, the two
principal groups which have exerted pressures in the most
recent years are parents and employers. For example, the
impetus for the 'back-to-the-basics' movement which built up
during the seventies could well have been due in part to
parental judgments and expectations, whilst employers have
repeatedly expressed concern about the mathematical qualifications of entrants into industry.

PARENTAL EXPECTATIONS

Reference has already been made to the importance of
role style and role expectation with respect to the teacher.
With the recently increased public emphasis on accountability
in the teaching profession, the feeling appears to have
arisen that the views of parents must be taken into account
to a greater degree than in the past. As Otte (1979) states
with respect to mathematics, 'Teachers are increasingly
forced by pupils and parents to justify their teaching ...
with respect to the selection of content and the relevance
of mathematics for the pupils' future life' (p.110). This
raises the question of what expectations parents have of
the teacher's role.

Musgrove and Taylor (1969) found that the aspect of the
role of the teacher identified by parents as most important
was, perhaps not surprisingly, the ability to teach. A
teacher's ability to teach would probably be judged by
parents largely in terms of their own children's success and
their reports of day-to-day classroom events, so that in
striving to meet their expectations, teachers are constantly
being judged by parents as well as by pupils.

The visibility of mathematics plays its part in this as
well. Bernstein (1975) distinguishes an 'open' context of
schooling where subject matter is less defined, and teaching
is less structured, from a 'closed' context. He further
claims that as the context of schooling moves from the 'open'
to the 'closed' end of a pedagogical spectrum, so the

23

pedagogy becomes more 'visible' or identifiable. By the
time secondary school is reached, pedagogy is moving towards
the closed end of the spectrum. Subjects tend to become more
rigidly demarcated and teaching methods more formal,
Bernstein suggests. This is seen to be due largely to the
fact that learning tends to become more abstract and bound
up in the context of examinations at this stage. Thus, while
their children are at primary school, parents may well find
it difficult to make judgments about pupils' performance in
some subjects because there tends to be a high degree of
subject integration, but the situation tends to change at
secondary level. However, as already noted, mathematics as
a subject stands alone. Even if used in other contexts, it
is taught as a separate discipline so that whatever the
scheme or syllabus followed, parents, at the very least, can
be aware as to whether or not their children know their
multiplication tables. Even where new mathematical content
is concerned, books have appeared specifically to help
parents understand the new mysteries being unfolded to their
children (e.g. *The New Mathematics for Parents*' by Heimer
and Newman, 1965). One would be hard put to find a similar
book on the subject of 'Environmental Studies'. It is
arguable, therefore, that even at primary level, mathematics
has a greater degree of 'visibility' than other disciplines
and, as a result, the teaching of it becomes more open to
criticism by parents than most other subjects.

EFFECTS OF PARENTAL INVOLVEMENT

The need for the involvement of parents in their child's
education is self-evident, and has been confirmed by recent
research.

A follow-up study by Ainsworth and Batten (1974) of 114
children from the Plowden primary survey (Great Britain,
D.E.S., 1967) through to secondary school reinforces the
need for parents to be informed. They found that the most
important parental characteristics linked with high pupil
achievement were 'ambition, literacy, and awareness' (p.123).
Surprisingly, the single variable most strongly related with
pupil success was the size of family from which the father
came, pupils whose fathers were 'only children' having the
highest likelihood of success. Cox (1979) suggests that the
main implication from a study carried out with a sample of
disadvantaged eleven-years-olds is that for intervention
procedures to be of any value to the pupils, it is vital to
gain the interest and co-operation of parents. In Newbold's
(1977) study at Banbury, he found that only about 50 per
cent of parents of low ability children showed any interest
in their progress at school.

It would seem, then, that for parents to have a positive
effect upon their children's academic attainment they must
have, together with the appropriate attitudes, an awareness

of the educational system and of how to manipulate it to the advantage of their children. As mentioned earlier a new factor in the life of secondary pupils is the opportunity for them to exercise choice and it is to be hoped that parents are made fully aware of the options open to their children. It would be reasonable to assume that parents have considerable influence on whatever choices are made.

Mathematics, as a discipline, may present something of an anomaly here. It may be that because some children show no particular ability in computation and number work generally, parents will think they are not mathematically able while, in fact, these children could have a high spatial visualization potential which would enable them to achieve well in other areas of mathematics. Conversely, parents may believe that their children are mathematically gifted because their number work is sound. Thus it could be that pupils with mathematical potential are not being encouraged or, at worst, are being discouraged by parents with respect to studying mathematics at a higher level while some who do not have adequate potential are being positively encouraged to do so. Selkirk's (1974) research reflects this anomaly. He deduced from his study of pupils who had opted to take mathematics at Advanced level that there were specific grounds for discouraging some candidates while encouraging others. For example, he noted that pupils who studied mathematics in strange combinations with other subjects (e.g. with Latin and history) achieved well and hence, he considered, more thought ought to be given to the encouragement of the study of mathematics outside the usual subject combinations. Certainly what is needed is for the mathematics teacher to help the parents to be aware of their child's potential and to guide them, if need be, to see that that potential is realized. 'Potential' is of course different from 'achievement' and it could be that parents tend to recognize only qualities of achievement.

THE EFFECTS OF CLASS ON PUPIL ACHIEVEMENT

Clearly, one way to counter uninformed judgments on the part of parents would be to ensure greater teacher-parent contact to allow for better communication between the two. Just over ten years ago, Morrison and McIntyre (1969) reported the amount of contact of this kind to be very poor. Ashton *et al.* (1975), however, found a higher level of parental interest than was evident from studies carried out in the past (e.g. Douglas, 1964) but their results suggest a steady decline in this interest as the school intake became more working class than middle class. While at one time the degree of parental contact may have been attributed to membership of a particular class, more recently awareness has grown of the lack of what could be called typical behaviour of either the working or middle class (Musgrave, 1979) although specific class variables such as linguistic code may act as an advantage or disadvantage in the pupil's learning (Bernstein, 1971).

In a study in which questionnaires were administered to 3400 pupils in 36 secondary schools, Witkin (1974) concludes that 'It does not appear that the social structure of schools and the experience of the children within them can be profitably described in terms of the class culture conflict model' (p.323). Witkin (1974) suggests, however, that the influence of the family is felt in the way family background may limit the extent to which a pupil uses the value systems, presented by the school, to good advantage. In the case of working class pupils, they may accept the values but not be socially articulate enough to benefit from them, while middle class pupils may choose to reject them altogether. Thus the influence of parents is subtly manifested and while it may be over-simplifying to reason in terms of class culture conflict, it would still seem to be the case that 'there are many parents who want their children to do well at school, but who have no idea of how to play this role of good parent' and who do not demonstrate the knowledge and attitudes appropriate to it (Musgrave, 1979, p.249).

The effects of class upon mathematics learning has been studied by Mellin-Olsen (1976) in Norway. He suggests that in order to succeed in further research on learning, the individual pupil's background must be taken into account. He stresses the importance of the need 'to know how he and his family experience school, how they define it, and what role it plays for them' (p.16). This is seen to be necessary in order to understand the conflicting message systems to which pupils may be exposed and presumably, if one gained such knowledge of family background, it might then become possible to understand how to help parents to play the role of the 'good parent' in the educational context.

EMPLOYERS' VIEWS

The other main group to set constraints on what mathematics teachers attempt are employers, and their voice has become a strident one in recent debates. In fact it could be argued that their apparent concern about the quality of mathematics teaching has been one of the instrumental factors in establishing the need for a national inquiry.

A study carried out by Bishop and McIntyre (1970) compared the opinions of employers and secondary teachers with regard to the content of secondary school mathematics and where emphasis should be placed in the teaching of the subject. Where the latter was concerned, six priorities were listed from which to choose:

(1)   its application to everyday life;
(2)   as a foundation for more advanced mathematics;

26

(3)   as an enjoyable and satisfying activity;
(4)   as a tool for use in a person's expected occupation;
(5)   as a foundation for scientific study;
(6)   for training children to think logically.
In all, 131 schools and 71 organizations from industry completed the questionnaire. While there was reasonable agreement overall, it is interesting to note that employers rated
'training children to think logically' as deserving greatest
emphasis, while teachers chose to place highest emphasis upon
mathematics 'as an enjoyable and satisfying activity'. The
employers' ratings in order were then 'application to life',
'the use of mathematics as a tool in an expected occupation'
and, ranked fourth, an emphasis on mathematics 'as an
enjoyable and satisfying activity'. Mathematics as 'a basis
for more advanced mathematics' and 'for scientific study'
were ranked almost equally, last.

Following their main priority of the enjoyable aspect
of mathematics, teachers placed 'application to everyday
life' next, while 'training children to think logically' was
placed third and then its 'use as a tool for a person's
expected occupation', while the final two were the same as
those of the employers, 'mathematics as a basis for further
study in science or mathematics'.

Clearly the most indicative outcome of this study is
the discrepancy between where each group sees the main
emphasis in teaching mathematics to lie. Teachers, in
placing the greatest importance on the enjoyable aspect of
learning mathematics, are probably indicating a belief that
pupils will not learn mathematics well if they do not find
it enjoyable and satisfying to do mathematics; also at a more
practical level pupils not enjoying their work may tend to be
bored and some may subsequently become disruptive. On the
other hand the highest rating given by industry to training
children to think logically would seem to indicate some faith
in the transference of mathematical mental processes and
skills to other areas of work or learning; certainly, pupils'
enjoyment of learning mathematics would seem to be a minimal
consideration to the members of industry who formed this
sample.

MATHEMATICS AND THE NEEDS OF INDUSTRY

Nisbet (1979) reports on a project called 'Understanding
British Industry' which involves people from industry who are
'committed to working with and through teachers' in order to
help them to develop confidence in transferring a knowledge
of industry to pupils (p.4). The two factors held to be
most important are that the initial training institution
must have people with some up-to-date experience of industry
and commerce, and that schools must provide a receptive
attitude for liaison between themselves and industry.

Fitzgerald (1978) suggests that it might be possible to develop hierarchies of tests that could correspond to the mathematical demands of particular sections of industry and that these could be used to measure a pupil's profile on leaving school, in order to discover where his potential for work in industry might lie. He argues that 'The variety of demands of different work situations is so wide that it seems unreasonable to expect pupils to be in peak form in all of them at any one time' but adds that skills once mastered can easily be revised (p.25).

SUMMARY

It seems to be the case that pressures on teachers from society have increased in recent years. With the growth of local, comprehensive schooling, with a rise in the number of working mothers, with the ever growing influence of media and with increasing worries and concerns over inflation and recession, demands on schools have grown and accountability has become a significant educational issue.

Mathematics, partly because of its perceived importance, and partly because of its 'visibility' in school, has been the focus for much of the accountability debate. The need clearly exists for teachers and schools to communicate with parents and employers, and to play their part in educating society about *realistic* judgments of aims, potential and achievement. At the same time teachers and schools need to take accountability seriously, and must reflect societal demands in their planning and teaching of mathematics courses. The role of the mathematics *department* in secondary schools seems critical here, as it is a more appropriate mediator and arbiter of these demands than is the individual class teacher.

Chapter Four
The Structure of the Teaching Profession

The external constraints upon teachers identified thus far have arisen from within the school or have been imposed by the demands and needs of society. A third powerful constraining force is exerted by the teaching profession itself, both through its structure and through the nature of the characterization of teaching as a profession.

THE NATURE OF TEACHING AS A PROFESSION

In his analysis of the role of the teacher, Wilson (1962) compares the demands made upon teachers with those made upon other professionals. The characteristics he selects to define such a role include the quality of the relationship of the professional with the client, the unquantifiable quality of service given, and the obvious moral commitment inherent in the role. Clearly, these characteristics can all be associated with the work that teachers do. However, whereas for other professions there is what Wilson (1962) describes as a 'definable expertise' through the application of the knowledge they have, this is not so precisely the case for teachers. Wilson (1962) states 'There is for the teacher what appears initially as a parallel - the objective body of mathematical, historical, musical or some other knowledge' (p.23). This analogy is not seen as an exact one, however, since teachers are not concerned with 'applying the rules of their expertise' but primarily with inculcating it. As a result, the type of service given is diffuse and the value judgments made are open to question. Consequently, the service offered does not receive proper recognition.

Added to the challenging of their value judgments in their professional role, the dilemma faced by teachers may be further compounded by other external factors. With respect to mathematics teaching in particular, for example, the sixties and seventies were times of great experiment and development. The confusion of new demands made from many directions increasingly involved the making of choices and the bringing about of change until, as Delaney (1977) suggests, mathematics teachers seemed 'to be sinking slowly under the weight of too many ideas and recommendations' (p.2).

29

With increased pressures of this kind, a sort of 'institu-tionalized guilt' was built up which may now be interpreted in hindsight as having eroded not only the confidence of mathematics teachers, but their autonomy as well. Thus it is not institutional factors alone which constrain the autonomy of the teacher as noted earlier but a variety of other external factors as well.

Grace (1978), in his study of teachers in urban schools says: 'The central meaning of autonomy for most of the teachers was a sense of freedom from interference (for whatever reason) *within* their immediate work situation: the classroom' (p.210-1). He found that while some teachers did not feel constrained by examination systems and viewed them as giving 'a sense of structure' to the educational enter-prise, others involved in innovation were more sceptical about the reality of their autonomy. Grace (1978) himself concludes that the possibility of innovation does exist at the level of the individual teacher, but within a framework of constraints, the chief of which are examination boards and the availability of resources. He suggests that only with the easing of such constraints can teacher autonomy 'have that reality which a present "liberal faith" now attributes to it' (p.213).

A particular illustration of the consequences of strong external constraints is offered by the increasing legislation in America to provide minimum-competency-based mathematics instruction. This has resulted in teachers treating the minimum established as their ultimate aim in pupil achieve-ment. Also, with a strong move towards individualization in mathematics teaching (Webb, 1980), the teacher's decision-making usually associated with planning (Clark and Yinger, 1980) has been usurped and consequently teachers are seen to be in danger of becoming 'de-skilled', an idea implicit in Morgan's (1977) study. Closer to home, Brown and McIntyre (1978) also studied teachers' responses to curricular innovation in Scotland. They found, with respect to innovations from outside the school, that there was a 'lack of evidence of any organised departmental influence on teachers' responses to innovations' and that this was related 'not to the autonomous decision of the teacher but to the general context within which the concerns of planner and teacher operate' (p.22). Meanwhile, in mathematics education in socialist countries, it would appear that the 'careful articulation of curricula, laid down centrally' is believed to prevent teachers from effectively bringing about any reforms of their own (Howson, 1980). Clearly it would appear that teachers in general find it difficult to operate as autonomous professionals, and that despite a desire to be members of one of the 'helping professions' (Edelman, 1974), where the priority of their role would be to help and to guide, in fact what happens is that they are thwarted in exercising their professional judgment in doing so.

These professional constraints are manifested particularly in the work situation in which teachers must operate. We have already seen, in the first chapter, how the within-institution relationships affect, through *social* mediation, what mathematics teachers can do. It is also the case that institutions have their own ways of restricting the possibilities of action by different teachers.

This is a situation that has been identified in connection with the teaching profession in other countries. Arfwedson (1976) discusses the dichotomy between the system of goals and the system of rules which is present in the organization of schools in Sweden. The goals of a school are not seen to be accompanied by sanctions since they are related to the pedagogical methods and attitudes adopted, whereas rules do carry sanctions. A teacher, for example, cannot disregard keeping to a timetable and recording pupils' attendance with impunity. However, Arfwedson (1976) points out that the power of the teacher is somewhat superficial since there would appear to be an inevitable conflict between rules and goals and although the teacher has apparent pedagogical freedom, the rules impinge strongly. 'On the one hand the teacher is a part of the hierarchical power-structure of the school organisation, on the other hand it is his duty to realise goals that are mainly democratic and anti-authoritarian' (pp. 141-2). Thus any desire on the part of the teacher to bring about change becomes inhibited or destroyed. Lortie (1975) refers to a similar situation in America when he states that 'Teachers have a built-in resistance to change because they believe that their work environment has never permitted them to show what they can really do' (p.235). Lortie's view is that as a result teachers often see the proposals for change made by others as 'frivolous' when they do not actually affect their working constraints. This type of conflict leads to complex feelings. For example, if Nash (1973) is to be believed, 'demoralized cynicism' is the 'occupational disease' of the profession and he suggests that the teacher's 'carefully preserved professional rights are more or less worthless' since in his view, 'No teacher can afford to act differently from the rest of the staff' (pp. 129-30). Francis (1975) strikes an equally pessimistic note when he states 'Conditions of work may account for teachers' caution, but they do not explain the full force of their cynicism which can be ferocious' (p, 151). Part of this cynicism he sees as arising from the fact that many teachers resent the position in which they find themselves, where they are expected to accept values and methods handed down by other people who are not themselves involved in actually applying them.

PRIMARY TEACHERS

According to Blyth (1965), primary teachers, because of the diffuseness of their role, can adopt something of a

defensive stance with respect to the hierarchy in the teaching profession, maintaining that it is much more difficult to
teach everything in a group of over thirty children than it
is merely to teach a subject. He suggests that status
differentiation with primary school staffs is difficult to
determine and that the salary structure is ill-adapted to
status-pattern. This is a situation which may have changed
as a result of the Burnham Report, although the basic grudge,
that a salary structure designed for secondary schools is
imposed upon primary teachers would sustain the primary
staff's impression that they are considered less important
than secondary staff (Blyth, 1965).

The recent D.E.S. (1978a) primary survey shows that of
the 5844 teachers in the sample, three-quarters were women.
Graduate status was found to be more usual among recently
qualified teachers, one-tenth of the total sample being
graduates, and two-fifths of those holding a Bachelor of
Education degree. Sixty per cent of the sample were on
Scale 2 or above, by far the largest proportion (35 per cent)
holding Scale 2 posts. Presumably, as the fact that new
appointments are decreasing was noted, the 'stability' that
follows will mean less likelihood of teachers receiving
promotion.

A more pertinent professional constraint on primary
teachers occurs because of their oft-quoted general lack of
mathematical expertise, an aspect on which their secondary
colleagues have always had a view. As Fielker (1979) says:
'... many secondary teachers - perhaps under their own
threats from employers and examinations - are only too
willing to tell their primary colleagues what to do.
Unfortunately this does nothing for their security, poses
more of a threat than not telling them, and does not even get
anything done. Mercifully, in practice this never happens
because secondary teachers cannot agree amongst themselves
about what they want!' (p.3).

One study (Bishop and McIntyre, 1969) which did consider what secondary teachers want, looked at the differences
between primary and secondary teachers' views on what
mathematics should be emphasized in primary schools. Although
it was undertaken ten years ago, the results are still of
interest. The major difference between the two groups was
that 'secondary teachers, being more concerned with a sound
foundation being laid for more advanced mathematics, stress
the understanding and efficient use of pure numbers. On the
other hand, the greater concern of the primary teachers with
application is shown by the higher importance they attach to
such things as map-reading, temperatures, percentages, area,
practical geometrical tasks and the graphical representation
of data'. Moreover 'Other general differences appear to be
that primary teachers are concerned with a wider range of
ideas than their secondary colleagues consider important for

the primary school; and that within the fields of verbal arithmetic problems and formal algebra they tend to think they should do more than secondary teachers consider necessary, or perhaps desirable' (p.3a).

With the growth of the work of the Assessment of Performance Unit, the judgments and abilities of primary teachers have become more of a topic for general educational discussion. The mathematical performances of 11-year-olds as shown by the first Primary Survey Report (Great Britain, D.E.S., 1980b) have become public, as well as professional, knowledge, and whilst the report does not overtly criticize primary teachers, phrases like 'sharp decline', 'grasp... tenuous', 'many pupils find - too abstract', 'proved diffi- cult' carry with them an 'end of term report' flavour which calls into question the primary teachers' competence. On the assumption that the tests were devised on the basis of commonly agreed content, one cannot avoid the conclusion that either primary teachers are trying to teach unlearnable ideas at that stage, or if the ideas are actually accessible to primary children, the teachers are failing to teach them successfully. The assumption and the conclusions are both clearly simplistic within the complex pattern of primary education in the U.K. Nevertheless, reports of this nature are likely to add to the already existing pressures exerted by the education profession on its members.

SECONDARY TEACHERS

It would seem to be the case that secondary teachers fare no better than their primary colleagues with respect to professional pressure. Some of the cynicism of teachers referred to earlier can be seen in the results of a study carried out by Hilsum and Start in 1971-2 to investigate promotion and careers of teachers (Hilsum and Start, 1974). Their sample involved 6722 teachers from 881 secondary schools from almost all Local Education Authorities. Teachers were asked to rank 12 factors they saw as favouring promotion and 12 factors they felt ought to favour promotion. With respect to the former, the first five factors in order of importance were (1) being a graduate; (2) being a specialist in a shortage subject; (3) social contacts; (4) conformity with advisers views; (5) good relations with the head. The factors they felt ought to favour promotion were (1) flexibility in teaching methods; (2) familiarity with new ideas; (3) ability to control pupils; (4) concern for pupils' welfare; and (5) having taught in a variety of schools. There is considerable discrepancy between these two sets of factors, with a clear personal emphasis in the first as compared with a more professional one in the second. This evidence does seem to suggest that teachers do not feel they are judged objectively when being considered for promotion.

The teachers' belief that the second most important factor affecting promotional chances is to be a specialist in a 'shortage subject' should augur well for mathematicians. However, the survey showed that mathematics teachers in the sample ranked seventeenth, according to subject taught, in gaining a Scale 2 post. By the time Scale 5 was reached they ranked ninth, as deputy heads they ranked equal fifth with English and geography and as heads, they ranked fourth. These statistics would not seem to bear out the teachers' belief that teaching a shortage subject favours promotion and this is some indication, however slight, that their views concerning promotion prospects may be misguided.

Within secondary schools promotion occurs within departments and clearly, the head of department plays a key role in determining such promotion. This process begins when probationary mathematics teachers take up their first post and a variety of demands is made upon them. In a pilot study carried out by Shuard (1973), sixteen mathematics heads of department indicated in questionnaires that they expected probationary mathematics teachers to undertake all the work of a full-time, qualified teacher of mathematics, with the exception of work connected with long-term planning. Bearing in mind the evidence quoted earlier in connection with the apparently poor performance of probationary mathematics teachers, this would suggest that the full role they are required to play places too great a strain upon them at this stage in their career. Cornelius's (1973) study also would appear to support this. Of 47 first-year mathematics graduate teachers in his sample, 31 referred to discipline as a major problem and the next most frequently mentioned problem (referred to by 18 teachers) was teaching children of low ability and mixed ability groups. A comment such as 'Discipline, especially with the less able who are uninterested in school, work and mathematics and generally disillusioned with life' is indicative of their problems while a further comment 'Inability to change anything ... policy is sent down and the people at the wrong end of the department have the dirty work to do' has a slightly cynical ring about it, reminiscent of Nash (1973) and Francis (1975) (Cornelius, 1973, p.13).

For the secondary mathematics teacher, the education profession has a well-founded institutional way of exerting its demands. Despite the fact that teachers may accept the examination system as necessary to the structure inherent in the secondary school, there is evidence to suggest that examinations exert a limiting effect on their teaching. H.M.I. (Great Britain, D.E.S., 1979b) voice some concern about this with respect to mathematics. They found in their survey that, 'Very frequently teachers considered that the need to cover examination syllabuses and the need for their pupils to cope with examination questions forced a restricted approach to the ideas embodied in the syllabus' (p.117). This reflects the fact that examination syllabuses tend to be controlled by university, higher and further education interests, which has resulted in recent demands by teachers for more involvement in the examining process.

SUMMARY

This chapter has focused on matters relevant to the structure of the teaching profession acting as a constraint upon mathematics teachers in particular, and reveals some of the problems and professional attitudes connected with teaching generally which may be assumed to be very much a part of a mathematics teacher's life. Wilson (1962) in his analysis of the profession points out that since the type of service given is so diffuse and the value judgments made are so open to question, the service offered is not given due regard. A different perspective is offered by Arfwedson (1976) and Lortie (1975) who both emphasize the features of the work place which prevent teachers from acting in a fully professional way, and Grace (1978) identifies the limited nature of their autonomy which results.

Primary teachers seem to be particularly vulnerable as a result of their relative lack of mathematical expertise, but secondary teachers cannot avoid professional pressures either because of the constraints placed on them by their higher and further education colleagues through the examination system.

Perhaps the resolution of these problems lies in increasing communication and mutual respect within the pro-fession. For example, Otte (1979) refers to 'the necessary continuous communication among different participants' which he views as a crucial contribution to the development of the professional life of mathematics teachers (Otte, 1979, p.127). Mutual professional respect is well exemplified by Trivett (1977) in an article entitled 'Which researchers help teachers do their job?' where he says 'Schools are very complex places with many complicated systems operating on and affecting every adult and child. Any teacher is right to pause before accepting the result of this or that simplistic finding of research; he is after all held accountable in more important and pressing ways for what his students do. It is not easy to sort out from all the possibilities exactly what for every moment guarantees any desired learning or behaviour effects' (p.42).

Such tolerance and understanding would go a long way to reducing the stresses caused by judgments made between professionals.

Chapter Five
The Effects of Initial Training of Teachers of Mathematics

The focus has until now been on the constraints, external to the teacher, which control and shape their possible actions. We now begin to consider those constraints which we think of as being internal to the teacher, in the sense that they are brought by the teacher to the teaching situation. The relevant research and conclusions are grouped in three chapters, this one, the next, Teacher Characteristics, and the third, In-service Training and Professional Development.

In this chapter we look at the effects which initial training produces on the student teachers, which lays the foundations of the knowledge, attitudes and perceptions brought by the students into teaching.

TEACHERS IN THE PRIMARY SECTOR

Since the James Report (Great Britain, D.E.S., 1972) there has been a movement towards making teaching an all-graduate profession. Judge (1975), himself a member of the James Committee, comments upon the 'poverty of thought' and critical discussion 'on the nature of the teacher and on the objectives and methods of teacher education' that were pervasive in the fifties and sixties (p.8). Referring to the Report, he goes on to say that 'The conviction that there should be a body of theoretical knowledge at once philosophically sound and applicable in good practice was stronger than the ability to say it'. Because of the visibility of the teaching profession, he suggests there is less agreement about what new entrants need to know and do than in any other profession, a point which relates strongly to that made in the last chapter concerning the nature of the teaching profession.

In a paper written three years after the James Report appeared, Shuard (1975) refers to the fact that "The new B.Ed. was intended as a professional degree which would improve on the Certificate' (Shuard, 1975, p.18). She goes on to say that the chances of this happening at that time were not good, and where mathematics in particular was concerned, it depended almost entirely on the qualifications of those recruited. Although prospects appeared not to be good, it was found that mathematics students were 'not worse qualified than the average student in other main subjects' (Shuard, 1975, p.18). In a survey carried out on 1975 entrants to colleges of education courses with mathematics as a main subject, returns from approximately 425 first-year students indicated they intended to stay on for a fourth year to take the B.Ed. in mathematics (Shuard, 1977). By the academic year 1976-7, in a similar survey it was found that numbers of main mathematics students had decreased by 14 per cent, but their qualifications had improved, with approximately 70 per cent having Advanced level mathematics (Shuard, 1978). Lumb (1974), in an earlier study, also investigated the initial mathematics qualifications of student teachers on entry to college. In his sample of 110 men and 186 women, 55.2 per cent had Ordinary level mathematics passes. They were tested on computation as well as some modern mathematics items and it was found that 86.4 per cent failed to score at all on the 'modern' mathematical questions. As an example of more general number work, in a question which involved placing five simple fractions in order of size, 76 per cent failed to do so. Lumb (1974) concluded that there should be a compulsory mathematics course for all college of education students.

Ray's (1975) research sheds more light on the social and attitudinal aspects of the students' mathematical training. He investigated factors which appeared to affect recruitment to main mathematics courses in colleges of education by studying a sample of 848 first-year entrants. The fact was noted that most students brought with them from schools a favourable attitude towards mathematics as a subject but not towards teaching methods used in conjunction with the sub-ject. More girls had dropped mathematics at school because they had seemed to be encouraged less to keep it up, and phrases such as 'not a girl's subject' frequently appeared on the questionnnaire. Only 73 per cent of all the sample had done Ordinary level mathematics. The criterion for getting onto the main mathematics course was a good Ordinary level pass (Grades 1-3) or an attempted Advanced level and, in the end only 19 per cent managed to do so. The main reason given for studying mathematics was an interest in the subject, but success in it was seldom referred to and other subjects were preferred more. It was of some concern to find that many students did not realize that they were qualified to do main mathematics. Apparently over a half of those qualified fell into this category which suggests a lack of vital information reaching students. Criticisms of the way they themselves had been taught mathematics at school included reference to 'humiliation', a heavy reliance on textbooks, a lack of individual attention and a lack of relevance to people and life. Among his conclusions, Ray (1975) suggests that a new approach to Advanced level mathematics is needed, stressing the encouragement of girls to study it at this level, and that the teaching of the sub-ject for this age-group, where method had had a particularly adverse effect, should be investigated.

Some attempt to explain what might happen to change college of education student attitudes to mathematics was made by Lumb and Child (1976) who tested those attitudes on entry to college and at the end of the first year. There was no substantial difference initially among those opting for first, middle or secondary schools. Although a somewhat limited study in design, it was found that only those who were going to teach in first schools showed a substantial improvement in attitude towards the subject. Clearly students' previous schooling and their college work have significant effects on the knowledge and attitudes they bring with them into the classroom.

TEACHERS IN THE SECONDARY SECTOR

Traditionally, requirements for qualifying as a teacher of mathematics at secondary level have had a strong academic flavour. Students generally have taken a degree in mathematics followed by one year of training to gain a Post-Graduate Certificate of Education, which has resulted in considerable emphasis on their mathematical knowledge rather than on their educational expertise. Some change has occurred due to the fact that teaching has moved towards becoming an all-graduate profession, whilst bringing with it a confusion in the variety of degrees that students may take. With respect to mathematics, students may take a Bachelor of Education with mathematics as their main subject or they may take a degree in mathematics and follow a Post-Graduate Certificate of Education course for one year as noted above. There also exists the possibility of following an Honours Degree in Mathematical Education. This means that initial training for teaching mathematics in secondary education provides teachers for whom different kinds of expertise may have been stressed in their initial training. In particular, the demand for high academic standards in colleges of education courses lays open the trap of 'too narrowly conceived academic standards in mathematics' which may be beyond the student's capacity to understand (Royal Society, 1976, p.18).

On the other hand, the problem arising from the view that appropriately high standards can only be attained in the academic courses offered at universities is that it may detract from the professional aspect of initial training (Royal Society, 1976, p.18). Thus at secondary level, there are three paths to becoming a graduate mathematics teacher each of which provides varying experience of this kind. A fourth type of mathematics training for secondary teachers entails following a certificate course with mathematics as a main subject.

The information from studies by Lumb and Child (1976)

and Ray (1975) quoted earlier concerning the initial training of college of education mathematics students is equally applicable to those intending to enter the secondary sector, where qualifications on entry are concerned. As already noted, Lumb and Child (1976) found that there was little difference in attitude towards mathematics amongst the three groups of entrants planning to teach in either first, middle or secondary schools and prospective secondary school teachers in the sample did not show a substantial improvement in attitudes towards the subject. Ray's (1975) study showed that those going to teach mathematics in the secondary sector would appear to be doing so because of an interest in it, with apparently little positive feeling in terms of success in, and liking for, the subject. This suggests that mathematics in some way qualified as a 'good' option for students in spite of their holding some adverse attitudes towards it.

Cornelius's (1973) findings with respect to graduate mathematics teachers' views of initial training suggest that, while finding teaching practice useful, 'courses had been too "general" or too "idealistic" and more discussion of discipline and problems of teaching low ability pupils would have been welcomed' (p.14). This emphasizes once again the essentially academic nature of the 'degree plus post-graduate training' route, in which the knowledge of the subject clearly predominates.

A study by Hoad (1974) sheds some light on the problems during teaching practice of graduate teachers taking a Post-Graduate Certificate in Education. Although not concerned particularly with mathematics student teachers his study shows how their subject might affect their position. He was concerned with the socialization of graduate student teachers in their schools and found that their social adjustment scores depended upon, amongst other things, their teaching subject. He considered that there was possibly a '"subject culture" transmitted from experienced teachers to newcomers' (p.159). If earlier discussion in this paper is considered this could mean that graduate mathematics students on teaching practice might feel in a privileged position as a result of being associated with a high status subject and the group identified with it in their school. However, Hoad (1974) also noted that school type was another factor affecting students' degree of socialization, and forming part of this are the pupils and the type of teaching within the school, factors which may well counter possible benefits to be gained from any kudos in being a mathematician. A further interesting aspect of his investigation was to examine students' adjustment in the light of their supervisors' role. His conclusion in this respect was that the relationship was unclear, but that the supervisor either provided impetus to the students' socialization or had a stifling effect on it.

There is no research evidence to suggest that graduate mathematics student teachers have more difficulties with discipline on teaching practice than other student teachers. However, there is some indication that more mathematics teachers have their probationary year extended than other subject teachers (Cockcroft submission, B18). Such information, if investigated appropriately, might provide valuable and useful information concerning the plight of mathematics probationers.

An interesting new approach to the study of discipline problems on teaching practice generally, which employs catastrophe theory, is reported by Preece (1977). It is suggested that the novelty of the approach 'hinges on the student's anxiety-induced failure to perceive accurately the level of disorder, and it does not depend upon his inability to act appropriately given accurate information' (p.23). If a further amount of instruction in appropriate behaviour is given the students, this only increases their anxiety. The conclusion is drawn that what is needed is to sensitize students to cues within the classroom situation to help them become more aware of these as potential problem sources.

The fact that there have been changes in teacher training techniques at graduate level has been noted by the Royal Society (1976). They suggest that there has been a marked effort to bring theory and practice closer together by integrating the theoretical discipline and making more of material from mathematical education, thus overcoming some of the difficulties which they believe students in mathematics usually find in this area. This is an attempt to overcome what the Germans call 'practice shock' which is seen as the phenomenon that exists where most graduates appear to lack practical teaching skills, a factor which they consider 'has hardly been taken into account in the reform of teacher training in the Federal Republic' (Mies et al. 1975, p.36). In this country, the Teacher Education Project based at Nottingham has been established in an attempt to come to grips with such problems (Kerry, 1977). The five main themes they have chosen to investigate with respect to courses leading to the Post-Graduate Certificate of Education are class management and control, mixed ability teaching, exceptional pupils, language across the curriculum and teaching skills. It would seem safe to assume that these themes represent a categorization of the main areas of concern for graduate student teachers. A further study of a similar nature is being undertaken at the University of Leicester.

It would appear that, while secondary mathematics teachers come to teaching with quite diferent kinds of experience depending upon the type of training they have undergone, the dichotomy that used to exist between those from a more 'academic' course and those from a more

41

'professional' one may not now be so clear-cut. At graduate
level, the problems arising from the traditional academic
emphasis may be being solved to some extent by new approaches
to training. However, there remains the problem of adjusting
the academic demands of the subject in professional degrees
like the B.Ed. to an appropriate level.

SUMMARY

The problem facing many primary teachers with respect to
the qualifications they bring to the teaching of mathematics
appears to be a lack of security in the mathematical knowlege they have, a concern identified by Ward (1979) in his
study. Begle (1979) has suggested that 'it seems to be
taken for granted that it is important for a teacher to have
a thorough understanding of the subject matter being taught'
(p.28). He quotes American studies (e.g. Popham, 1971)
which indicate that 'this belief needs drastic modification
and in fact suggests that once a teacher reaches a certain
level of understanding of the subject matter, then further
understanding contributes nothing to student achievement'
(Begle, 1979, p.51). It may be, however, that the desirable
optimum of mathematical knowlege for primary teachers in this
country has yet to be achieved.

At secondary level, it would seem that the problem is
reversed and the lack of balance appears to arise in the
area of professional expertise. The 'academic' nature of new
degrees, as well as the Post-Graduate Certificate of
Education, which has characterized these forms of initial
training has led to suggestions that emphasis on professional
expertise may have suffered. The Royal Society Mathematical
Sub-Committee (1976) refers to the complex nature of the
demands made upon teachers of mathematics when they state
that: 'it is not clear how far the special knowledge and
training of the totality of mathematics teachers is adequate
to the demands which the teaching of mathematics, rather
than teaching in general, places upon them' (Royal Society,
1976, p.3). Thus for initial training at both primary and
secondary level, there would seem to be a need to bring into
balance appropriate mathematical knowledge with the appropriate professional skills to impart that knowledge
successfully.

Chapter Six
Teacher Characteristics

Referring to teachers of mathematics, Griffiths and Howson (1974) write 'Apart from technical competence, a good teacher will also have dedication, unselfishness and a wish to make his pupils *better*' (p.67) (their italics). Although the descriptive 'better' may beg some philosophical examination the authors link the production of better pupils with developing 'intellectual distinction' in teachers of mathematics so that, ultimately, they will not be satisfied 'with the production of competent dullards' as a result of their labours (p.67). No doubt other subject teachers would wish to embrace similar educational aims but the idea that mathematics teachers in particular should wish to pursue such a goal may lie in the fact that it is hoped that their perception of their subject will be of a broader, richer nature than in the past.

At primary level, teachers' perceptions of mathematics as a subject are likely to be determined by their limited mathematical background, as suggested in the last chapter. Ward (1979) bears this out in presenting the results of a survey undertaken for the Schools Council. He reports that in 1974 the primary teacher's main handicap with respect to mathematics was a lack of mathematics education; less than 60 per cent of teachers in his sample of 40 schools had Ordinary level passes and less than 5 per cent Advanced level passes in mathematics. The suggestion is made that mathematics can suffer more than any other subject from poor teaching because of the linearity of the subject. Teachers may tend to follow through topics in a step-by-step approach, which possibly lacks breadth and depth and does not make use of appropriate concrete experience, because they are not confident enough in what they are doing to deviate from the narrow factual path. At secondary level, the dangers of the 'too narrowly conceived academic standards' which may be perceived as inherent within the subject have also been identified (Royal Society, 1976, p.18) together with their potential effect on the teaching of the subject.

Teachers' perceptions of mathematics clearly are a vital constraint in the complex classroom situation in which they work. These perceptions inevitably interact with, and affect

43

other, teacher characteristics which further constrain the teaching/learning situation, and outcomes in terms of mathematical learning. While considerable research has been carried out on the characteristics of teachers, studies most pertinent to our considerations seem to arise within three areas of concern:

(1) teacher attitudes and their effect upon pupils:
(2) teacher expectations;
(3) the effects of teachers' perceptions of the mathematical performance of girls.

ATTITUDES OF MATHEMATICS TEACHERS

Begle (1979) has collated the results of American research into mathematics education and included in this are studies investigating the characteristics of teachers of mathematics. He draws on data from the National Longitudinal Study of Mathematical Abilities (NLSMA) carried out over a period of five years and involving over 100,000 pupils and their teachers.

In investigating attitudes, seven different variables were extracted from questionnaires sent to teachers involved. The variables were:

(1) theoretical orientation (whether teachers placed emphasis on teaching for understanding or rote learning);
(2) concern for pupils;
(3) involvement in teaching;
(4) non-authoritarian orientation;
(5) like versus dislike of mathematics;
(6) creative versus rote view of mathematics;
(7) need for approval.

Each of these seven variables was found to have a positive main effect on pupil achievement in mathematics (Begle, 1979, p.46). On superficial examination, positive attitudes linked with these variables in mathematics teachers, might well be expected to have a marked effect on increasing pupil achievement. Although Begle himself suggests that this is not the case, the findings of the NLSMA bear some discussion.

Firstly, where the teacher's theoretical orientation is towards the development of understanding as opposed to rote learning, the NLSMA found that greater pupil achievement resulted. Similarly, a greater satisfaction and interest in teaching, a greater liking of mathematics and a belief that learning mathematics is a creative process all appear to relate to higher pupil achievement. Possibly one of the more interesting results is the indication that the less empathy and concern on the part of the teacher for the social and emotional aspects of the pupils, the greater the pupils' success in mathematics. This result may seem incongruous with respect to the situation in this country and the trend towards a greater concern for interpersonal relationships

44

within the classroom (e.g. Hargreaves, 1972; Keddie,1971). However, it is supported by Bennett (1976) to some degree, when he found teaching style to be far more important than teacher's personality where pupils' progress was concerned, whatever the subject. Thus, attributes such as 'warmth' sometimes judged to be desirable in teachers would appear to be of less importance to mathematical achievement than a commitment to teaching for understanding with an emphasis on the creative nature of the subject, according to American studies.

A further interesting correlation with higher mathematical achievement which has been suggested is a non-authoritarian attitude on the part of the teacher. This is seen as an approach opposed to the enforcement of strict discipline. Bossert (1979) notes that 'task structure influences the degree to which teacher and pupil behaviour is public and activities depend on teacher control' and, as a result, different kinds of task demand different kinds of teacher control (p.62). If mathematics is viewed in a creative light, as is suggested is desirable, then it would probably involve a considerable degree of activity on the part of the pupils and strict, formal discipline would be difficult to maintain. If a creative approach to the teaching of the subject correlates with greater pupil achievement, it would seem logical that a non-authoritarian approach should also prevail. This is substantiated to a degree once again by Bennett (1976). While he suggests that there appear to be few differences in pupils' performance attributable to teacher type, due to the difficulty in defining teacher types, he acknowledges that indirect teaching can lead to increased gains in pupil learning in creative areas. He states that there is some support for this being the case with respect to the teaching of arithmetic concepts, in particular at primary level.

Finally, the relationship between a teacher's need for approval and high pupil achievement would seem to be fairly logical. Teachers are aware that they are most likely to be judged in terms of pupil achievement and a teacher who needs approval will strive hard to see that pupils do achieve well.

Begle (1979), however, is sceptical of all of these results and notes that, on examination, none of the variables had a very strong correlation with pupil achievement and, hence, he concludes, no strong influence on teacher effectiveness. There appeared to be differences in the effects of some variables depending upon whether pupils studied conventional or modern mathematics courses (although he does not identify which variables) and he notes that there were no differences between boys and girls where the distribution of effectiveness of variables was concerned.

Begle's scepticism led him to undertake a further analysis of the data gathered by the NLSMA (Begle and Geeslin, ED O84 13O). The main conclusion drawn was that 'significant relationships between teacher variables and effectiveness scores were not frequent, appearing in fewer than 30 per cent of the possible cases' (Begle, 1979, p.50). They found affective variables to have a stronger effect than background variables (e.g. teacher's sex or marital status) and that the stronger affective variables differed depending upon the age of pupils concerned. For example, at the 11-year-old level, the stressing of the creative aspect of mathematics by the teacher was found to have the greatest effect upon pupil achievement, while at the 16-year-old level the satisfaction of the teacher's need for approval correlated most strongly with pupil achievement.

However, as Begle (1979) suggests, concurring with evidence gathered by Rosenshine (1971), 'The very concept of the effectiveness of a teacher may not be valid' (p.37). Teacher effectiveness is a quality which may vary over a period of time, thus rendering quantitative studies in this respect open to question. It would appear that there are no promising indications of relationships between teacher characteristics and pupil achievement and that 'our attempts to improve mathematics education would not profit from further studies of teachers and their characteristics' (p.55).

Brophy and Good (1974) also report studies which attempted to relate teacher attitudes to pupil characteristics. The attitudes concerned were attachment, indifference, concern and rejection and, as noted by the authors, once exhibited towards a pupil, such attitudes can lead to the beginning of self-fulfilling prophecies. The studies showed that pupils to whom teachers exhibited attachment were high achievers and conformed to a pattern of desirable classroom behaviour at the same time apparently being shown little evidence of overt favouritism on the part of the teacher. Pupils shown an indifferent attitude by the teacher were characterized by passivity and inconspicuousness. Even when perceived by teachers as unhappy or shy or nervous, one such study showed that these pupils still did not elicit the concern of the teacher but that the teachers appeared to be 'truly indifferent' to them (Brophy and Good, 1974, p.160). Students to whom teachers showed an attitude of concern were given much of the teacher's time in effort and help. The pupils to whom an attitude of rejection was shown, superficially appeared to be little different from the 'concern' pupils, but gained the teacher's attention primarily in the course of being disciplined in the classroom.

Nash (1972) carried out an investigation into teacher attitudes involving eight teachers and 236 pupils. He used the repertory grid technique, obtaining bi-polar constructs from the teachers involved, choosing the eight most highly

46

ranked constructs and converting them to a rating scale, then gaining a rank order of all pupils in each teacher's class. Pupils were observed as objectively as possible and their behaviour then reinterpreted in the light of the teachers' perceptions of them. Discussion of the results indicates the difficulty of understanding why teachers' attitudes to particular pupils are as they are and how idiosyncratic they may be. For example, it was found that the class teacher's estimate of pupils' ability was not related to the pupil's social class, as is sometimes assumed. However, the importance of the teacher's attitudes and perceptions upon the achievement of pupils is established.

TEACHERS' EXPECTATIONS OF PUPILS

The notion of the self-fulfilling prophecy with respect to teachers' judgments of pupils' ability was referred to earlier (see p.18). Good and Brophy (1978) comment critic- ally upon the study carried out in America by Rosenthal and Jacobson (1968) where attempts were made to manipulate teacher expectations by attributing false I.Q. scores to pupils. The intent was to study the outcome in terms of pupils' achievement and to discover the extent to which the self-fulfilling prophecy syndrome in fact existed. It was found that where a falsely high rating of pupils' ability was given to teachers, the pupils achieved more than would have been expected from their actual I.Q. score. Good and Brophy (1978) argue that this situation arose because of the 'credibility of the source' of the information about the pupils' potential, i.e. who it was that identified the high and low achievers. In what they call 'naturalistic' studies these expectations can be related to 'differential teacher behaviour' (p.68). Since in the normal course of events, teachers must and will make inferences of this kind, it is suggested that their perceptions and expectations could be guided by making them aware of what they probably may expect before they reach the classroom. In this way, they could be helped to make as well informed inferences as possible. They refer to a model in which the effects of teacher expectations are presented as outcomes of a series of cause-and-effect relationships. For example, it is suggested in the second stage of the model, that teachers begin to treat pupils differently according to their perceptions of them after an initial period of contact.

In his study 'Classrooms Observed', in which he studied a sample of pupils from five primary schools and followed them through to a single comprehensive school, Nash (1973) comments that in any classroom there exists a 'community of knowledge' held by pupils and teachers about the relative ability of each member of the class (Nash, 1973, p.90). The importance of the teacher's perception of the child and the effect of this on the child is highlighted. Where children are perceived unfavourably by the teacher, he suggests they will in turn develop unfavourable self-concepts and 'these will be reflected in the low class positions these children will believe themselves to have' (Nash, 1973, p.91). The converse of this statement was also believed to be true, that is, those who were perceived favourably by the teacher will consider themselves to have a good position in the class. These findings clearly reinforce those reported in the previous section on teacher attitudes.

Good and Brophy (1978) make the further point that the self-concept of pupils results from 'their early experience and the subtle but systematic opportunities and rewards they experienced', and they state further, 'Children are not born with inadequate self-concepts. Self-worth is learned in interaction with others' (pp. 82-3). The classroom situation in which a large proportion of this interaction occurs is characterized by Jackson (1968) as conveying a threefold lesson which the children have to learn in order to survive and develop their self image. They must learn (1) to live in a crowd; (2) to adapt to the fact that they are under conditions of constant evaluation both by teacher and their peers; (3) to understand the condition of power that exists within the classroom, with the teacher in authority and wielding the power. Mathematics classrooms could well be extreme examples of these three factors to which children must accommodate.

THE EFFECTS OF TEACHERS' PERCEPTIONS OF THE MATHEMATICAL PERFORMANCE OF GIRLS.

A considerable amount of attention has been directed towards the effect of the teacher's perception of pupils on pupil achievement, as noted earlier. One such study in Scotland involving a sample of 16 schools found, for example, that while assessments of achievement were closely related to objective measures, these assessments were 'to some extent affected by the teachers' perception of other characteristics of their pupils' (Morrison, McIntyre and Sutherland, 1965, p.318).

An interesting factor emerging from this study and confirmed in a later study (Morrison, McIntyre and Sutherland, 1966) arose in connection with teachers' perceptions of girls' achievement. In the earlier study they found that teachers, whether male teachers or single or married female teachers, tended to make 'a more general evaluation of girls than boys' (Morrison, McIntyre and Sutherland 1965, p.319). They were less analytic in their approach to rating the girls and tended to associate attainment with traits such as 'sociability and leadership'. In particular they associated girls' attainment in arithmetic more with 'good behaviour' than they did in the case of boys. The later study confirmed this tendency and here the conclusions drawn were that 'Teachers appear to make a more complete assessment in terms of one or two major dimensions of girls than they do of boys; and they vary much less in the qualities which they look for in girls' (Morrison, McIntyre and Sutherland, 1966, p.279). The 'ideal girl' would appear to be the same whatever her social class or background while there is greater variation with respect to the 'ideal boy' depending upon possibilities or limitations of the individual's environment.

Of all the considerations of pupils as a constraint upon teachers thus far, perhaps the latter point concerning teachers' perception of girls is one of the most important with respect to the teaching of mathematics. There has been considerable interest in the fact that fewer girls have chosen to study mathematics at a higher level in secondary education and in particular there appears to have been a 'tremendous shift' away from studying it at degree level (Cockcroft submission, J52). Concern has reached the point where one LEA (Essex) has set up a Committee for Girls and Mathematics as an aspect of the county's in-service training programme. In this country, there were recently at least two research projects investigating the general problem of low numbers of girls studying mathematics and bias in the curriculum towards favouring boys (Berrill and Wallis, 1976; Preece, 1979).

There seems little reason to expect that girls and boys should necessarily differ in their potential for mathematics. Certainly at primary level, evidence from the U.K. (Great Britain, D.E.S., 1978a) and from America (Fennema, 1979) shows little significant difference between boys' and girls' mathematics scores. Where any differences do occur at secondary level there is more concern about the differential teaching they have perhaps received. For example in connection with visuo-spatial ability, which some people claim is at the heart of mathematical ability, Berrill and Wallis (1976) suggest that the kind of experience boys enjoy at pre-school and primary age gives them an advantage over girls. It is suggested elsewhere that 'Traditional time-tabling patterns in many junior schools tend to provide only boys with three-dimensional modelling and construction' thus adding to the experience of this activity they have out of school and often depriving girls of their only chance to obtain it (Cockcroft submission, K19).

Perceptions become active discriminations through, for example, the illustrations used for various mathematical problems e.g. where boys buy planes and trains, while girls are given a doll and a doll's house (Berrill and Wallis, 1976). The effect of such perceptions has not been widely studied here but is considered highly significant in the U.S.A. For example Luchins (1979) observed that high school counsellors often discourage girls from pursuing mathematics and preparing for quantitative careers because the counsellors do not think that these activities provide opportunities for girls. Armstrong (1980) says 'It is the active encourage-

ment of parents, teachers and counsellors which seem to
affect participation (in high school mathematics courses)'
(p.30). Burton (1978) even suggests specific techniques for
teachers to assist girls in overcoming a negative self-image
in mathematics: (a) Do not reinforce sex stereotypes;
(b) Don't expect one sex or the other to excel in a given
area; (c) Encourage the development of spatial skills;
(d) Be a role model for mathematical learning; (e) Invite
guest speakers who are good role models; (f) Never accept
less than a child's best work; (g) Explain the career
relevance of mathematics to students.

SUMMARY

    The studies reported here have attempted to illuminate
the effects of the perceptions and attitudes of teachers upon
pupils in their charge. There have been three different foci
for these studies - teacher attitudes, teacher expectations
and the effects of perceptions on girls' mathematical performance. In general the results complement each other, and
Roberts (1971) describes the general effect when she writes,
'The teacher must be made aware of the potency of his
expectations. Research shows that very, very simple acts on
the part of teachers result in astonishing behavioural
changes in students' (Roberts, 1971, p.174). A study by
McKeachie et al. (1955) is quoted in which merely allowing
pupils to comment, in writing, on tests which they were
taking, resulted in a tendency for them to achieve higher
scores. This is interpreted as arising from a release from
anxiety on the part of the pupils. 'Student anxiety
evidently centers on their helplessness in relation to the
teacher's power. Freedom to make comments seems to relieve
anxiety about the possible arbitrary or punitive use of
power' (Roberts, 1971, p.174). It is clearly important for
teachers to be aware of how their perceptions of pupils can
lead to the conscious or unconscious exercise of such power.
In the mathematics classroom where overt judgments of the
abilities of pupils are made more frequently and more
publicly than in other areas, this is especially the case.

Chapter Seven
In-service Training and Professional Development

Hoyle (1979) suggests that the professional development of
teachers includes not only activities of an in-service
training kind but extends beyond to teacher participation in
a variety of other kinds of activities. Fletcher (1975)
identifies the particular problems involved for teachers
of mathematics when he writes:

'i) We have to take the teachers we have and teach them more
     about mathematics and ways of teaching it,

ii) we have to improve initial training, so that we do not
     have such a difficult task of in-service training in
     the future' (p.204).

He acknowledges these problems as important and goes on to
suggest that consideration must be given to the quality of
people doing the job of teaching mathematics, that they must
be 'better in the sense of more fully informed, wiser and
more adaptable human beings' (p.204). Fletcher then leads on
to what is his real concern, a discussion of the professional
status of the teachers of mathematics, drawing the conclusion
that they must not only be teachers but mathematicians, i.e.
members of the mathematical community. At the same time,
Howson (1975) refers to mathematics teachers shifting their
position 'towards that of the full professional' through
involvement with curriculum development but notes that 'this
must be founded on and integrated with, an effective in-
service programme' (p.278). Thus in-service training for
mathematics teachers is seen to be the basis for a variety
of kinds of activity which can help to extend them
professionally.

IN-SERVICE TRAINING

     In-service training has been identified as serving needs
at three different levels: that of (1) the individual
teacher; (2) a group within a school (e.g. a mathematics
department) and (3) the school as a whole (Great Britain,
D.E.S., 1978b).

     A main source of, and support for, in-service training
since the rise of the curriculum development movement has
been teachers' centres. Their function, as anticipated by
the Schools Council in 1967, was somewhat loosely described
in terms of providing a setting for discussion by teachers

and generally to focus local interest on curriculum development (Schools Council Working Paper 10, 1967). Most
teachers' centres aim at providing support for general subjects but some are specifically mathematics centres. Even as
specialist mathematics centres, their function would seem to
be less highly defined than the *Instituts de Recherche sur
l'Enseignements des Mathematiques* in France where part of
their role is to gather and disseminate research concerned
with mathematics education as well as to contribute to the
initial training of mathematics teachers (Revuz, 1978). The
success of teachers' centres in the past has depended upon
the numbers of teachers attending them; however, we are
reminded that in-service training 'is a voluntary professional activity which depends for its success upon the good-
will of teachers' (Great Britian, D.E.S., 1978b) and the
numbers of teachers using the centres may not always be
ideally what might be expected (p.3). Howson (1975) notes
one such centre, concerned to a large extent with mathematics and described as 'particularly active', which aims at
involving only 10 per cent of the total number of local
teachers (p.286). The D.E.S. (Great Britain, 1978b) booklet
concerned with in-service training makes a plea, as does
Howson (1975), for such training to become school-based.
This has the advantage of making the involvement (or non-
involvement) of all staff in relevant developmental work
open, and it becomes possible for all to participate together.
As this change in emphasis of approach takes place, clearly
the emphasis in the role of the teachers' centres will
change with it.

At primary level, the need for in-service training in
mathematics has been identified by the teachers themselves
(Ward, 1979), and the degree of the need may best be
illustrated by the fact that not only do they ask for further
in-service training courses but some appear to favour a
strongly centralized direction, if necessary something like
a 'national manual' (Ward, 1979). This is in spite of the
fact that 88 per cent of **pri**mary schools in the National
Survey of Primary Education found that the subject was
supported by either guidelines or a scheme of work, which was
a higher percentage than any other subject (Great Britain,
D.E.S., 1978a) Ward (1979) interprets this as evidence of
'widespread uncertainty', (p.57) a matter which to some
extent has been taken cognizance of with the publication of
Mathematics 5-11: A handbook of suggestions (Great Britian,
D.E.S., 1979a). However, the initial training of primary
teachers would appear to place them in a position of having
to teach mathematics when many apparently have not only a
poor attitude towards the subject but lack confidence in
their ability to teach it as well, as noted earlier.
Clearly, it is this situation that has led them to identify
their need for further training and development in the
processes of mathematics education.

Other evidence suggests that 'Finance should be forth-coming to improve the level of INSET which can be undertaken', since it is held that 'INSET programmes are generally inadequate ' to meet the needs and demands of primary schools (Cockcroft submission, J61). As in-service training becomes more school-based, at primary level such work could be led by the mathematics co-ordinator, where schools have them, or by a mathematics adviser or advisory teacher. However, there is doubt as to whether an adequate number of mathematics advisers exists throughout the Local Education Authorities to cope with the needs envisaged (Cockcroft submission, J61). Straker (1978) argues that because many mathematics advisers have other duties, there is a case for general primary advisers becoming more skilled in judging and guiding mathematical activities, ultimately, as she says, to compensate for the teachers' inadequacies in mathematics, 'and many primary teachers do feel inadequate' (p.13).

At secondary level heads of mathematics departments would be expected to take the role of leading in staff developmental work. This is a point raised by Neill (1979) in a paper discussing his appointment by the University of Durham School of Education to a post for the promotion of in-service training for teachers of mathematics to 11 to 18-year-olds, in the local area. He suggests that heads of department need training since the major responsibility for in-service training in schools will lie with them and he considers that this should be accepted as an integral part of their job. Neill's own experience with teachers has led him to believe that they are more concerned with organizational matters (e.g. the teaching of mixed ability groups), rather than mathematical matters, which is where he believes their needs lie. Rather than after-school courses which apparently are not successful, Neill advocates 'a major contractual commitment to INSET' on the part of teachers which would involve something such as the award of a diploma or gaining time off for study (p.7).

Where in-service training may involve the whole staff of a school, it could have some advantages from the point of view of acting as a unifying agent among different teacher interests or various subject departments. If, for example, a course of discussions were to be held in a school to consider the implications of the Bullock Report (Great Britain, D.E.S., 1975) across the curriculum, mathematicians might be made more aware of problems shared by other colleagues in how to use language appropriately and effectively in the teaching of their subject. This would take mathematics teachers beyond the bounds of their own discipline, make them more aware and better informed, and be a contributory step towards broader professional development.

Otte (1979) describes teacher education and the teaching of mathematics as being faced at present with demands from two different directions in development: 'on the one hand, the trend towards a far more accurate and differentiated special knowledge and specialization; on the other hand, the trend towards a more active participation of new groups of people and the inclusion of a growing number of areas of experience in discussion, analysis and decision making' (p.127). His analysis of the situation in Germany reflects also what is happening in the U.K. at the present time. Mathematics teachers now have available to them award-bearing courses, the purpose of which is to deepen their understanding and knowledge of their subject. Such courses are offered, for example, by the Open University and Schools and Institutes of Education. It is now also possible for primary teachers to return to college for a year to 're-train' in order to make mathematics their special subject. It would seem that there may be some optimum with regard to a teacher's mathematical knowledge and their effectiveness in teaching it, bearing in mind the evidence of Begle (1978). This suggests the possible need for some system of counseling for teachers who express a wish to undertake further professional training of this kind, and who may need guidance in the kind of mathematics they perhaps should study.

A large part of the professional development of mathematics teachers takes place in association with curriculum development projects at national, local or school level (Hoyle, 1979). This usually involves the mathematics teacher with educationalists of wider interests, who will help to bring psychological and sociological considerations to bear on the curricular work at hand, and ideally teachers from other disciplines as well. It may also involve others from outside education, for example from industry (Griffiths and Howson, 1974; Howson, 1975). Thus involvement in curriculum development can lead to mathematics teachers being drawn into the wider discourse considered desirable for their further professional development.

A final way in which this development may be extended is for teachers of mathematics to take part in research. It would be difficult to conclude that the teachers who took part in the illuminative study of four mathematics class-rooms by Yates (1978) were not affected professionally, from a variety of points of view, including how they viewed their subject, their pupils and their own ability to teach. As Cooper and Ebbutt (1974) state, in discussing their experiences as teachers participating in an action-research project (the Ford Foundation Teaching Project), 'the Project has made the teachers here think deeply about their methods and techniques. We feel that this and the discussions which have followed such thoughts have been very valuable' (p.70).

Within the mathematics teaching area, the Research Group of
the Association of Teachers of Mathematics (1973) has shown
the feasibility and value of this kind of in-service
activity, not only for the participants but also for anyone
contemplating the role of research in education. Brookes
(1978) states, 'The more questions are asked by teachers of
those thought to be expert, the more it becomes clear that
there is a lack of an adequate means of conducting appropriate dialogues between them' (p.x). It may be possible,
given the new directions being taken in research in mathematical education, with the emphasis shifting from quantitative to qualitative approaches (Kallos and Lundgren, 1976),
for teachers to become even more directly involved in
research programmes. It would appear to be a shift in
emphasis that may allow the development of the means of
enabling appropriate dialogue to take place, thus helping to
bridge the gap between teacher and researcher, between
practice and theory, and hence add to the professionalism of
both. Cooper and Ebbutt (1974) say in their conclusion, 'We
are pleased that this project has brought research workers
into the school - it seems to have helped them to understand
our problems, and helped us to understand theirs' (p.71).

SUMMARY

The picture presented here is one of differing needs and
demands on the part of teachers for continued professional
development. In-service training is seen to be the basis for
any further development of this kind and the paramount concern at primary level is, once again, that teachers must
deepen their own mathematical understanding. Professional
development can only come from increased confidence of this
nature.

At secondary level there would appear to be potential
conflict as to where the greatest need lies in the in-service
training and professional development of mathematics teachers.
During the early days of modern mathematics, the concern
would have been seen somewhat simplistically as one of
increasing the teacher's mathematical knowledge. While this
may contribute towards teachers becoming closer to the
mathematical community (identified by Fletcher (1975) as
being desirable), it is not sufficient to provide the full
professional status advocated by Howson (1975). Otte (1979)
or, indeed, by Fletcher (1975) himself. The conflict would
appear to be one of achieving balance between the extension
of mathematical knowledge and the extension of professional
knowledge which comes with activities such as research and
development work. There are obvious dangers inherent in
adopting one approach to the exclusion of the other where,
for example, the specialist mathematician becomes just a
purveyor of mathematics.

It is clear that the balance to be achieved is one of

individual need. While many mathematics teachers may have sufficient mathematical knowledge, they may well lack the broader perspective provided by adequate knowledge of class-room processes, or the situation may be the reverse. There is no doubt that to be a member of the mathematical community *and* to bring a full range of professional skills to the teaching of the subject must pose a considerable demand upon any mathematics teacher. It is difficult to assess the realism of such a demand, but it argues strongly that more individual guidance needs to be given in the professional development undertaken by mathematics teachers.

Chapter Eight
Some General Conclusions

Research into the social context of mathematics teaching is at a very formative stage and it is risky to attempt to write a chapter of conclusions after only a year's analysis of such research and studies as we have been able to find. Nevertheless it is important to attempt to pull together various ideas and to examine their implications if this analysis is to be of help to those involved in mathematics education, at whatever level. The present chapter, then, attempts to summarize our understanding of the social context of mathematics teaching, which lies behind our recommendations to the Cockcroft Committee (see Appendix, p.79).

External constraints: the non-autonomous teacher

We have come to understand more about the position of the teacher of mathematics within our educational system. This understanding certainly exposes what Maclure (1968) calls 'the myth' of the autonomous teacher. It is our feeling that this myth needs to be publicly exploded in order that due focus can be placed on the roles of others within the education system who control, knowingly or not, the conditions in which the individual teacher of mathematics operates, and in which children do or do not learn mathematics.

The idea of the 'autonomous teacher' guides much of our system's thinking about planning, teacher education and resource allocation, but the teacher as the slave to everyone else's 'good' ideas is more akin to the perceived 'self' of many teachers. It may be preferable to think of the ideal mathematics teacher as a creative and resourceful curriculum developer, skillfully combining the various ideas learnt from courses, books and those in advisory roles, to create a satisfactory mathematical education for individual pupils. However the teachers' reality may be more unpleasant – harrassed cynics feeling utterly frustrated in their genuine attempts to help their pupils, by the rigid and unhelpful conditions created by others, in which they must operate.

Much stress is engendered when teachers enter a system where they find themselves distanced from the top of the hierarchy where rules are made, and which they are expected

57

to obey, even though they may disagree with them (Francis, 1975). As Nash (1973) puts it, they cannot afford to act differently from the rest of the staff and implementing a system with which they do not agree clearly may cause teachers distress. For example, a probationary mathematics teacher may be faced with teaching a mixed ability class at secondary level (a factor rated highly on the list of problems of mathematics probationers in Cornelius's (1973) study). Every instinct of such a teacher may be to divide the pupils into ability groups within the class and to approach the teaching of each group in entirely different ways, but this could go against the ethos of the department or school completely, and the personal dilemma which ensues can be great. Clearly no teacher can be autonomous, and the roles that others play must be taken into account when considering that of the individual teacher.

LEADERSHIP ROLES

Of crucial importance seems to be the idea of *leadership* and how it is handled within both primary and secondary schools.

It is clear that primary head teachers have a great deal of potential influence in their schools as leaders. However, the combination of evidence from schools involved in the study concerning teachers' opinions about the aims of primary education (Ashton *et al.*, 1975) and that of the D.E.S. (Great Britain, 1978a) primary survey, suggests that some primary head teachers need guidance in how to take a more positive leadership role than heretofore. For example, the fact that teachers tended to adopt a more traditional role in schools where few direct, formal meetings between head and staff took place may not necessarily be a bad thing in itself. On the other hand, it may not augur well for the teaching of mathematics. If primary teachers withdraw in isolation to their classrooms, and if at the same time they are not con-fident in teaching mathematics, there is a strong possibility that such teachers will tend to limit mathematical content to computational skills only, perhaps teaching largely by rote, with little emphasis on the application of skills and concepts or on the use of concrete apparatus to help pupils to learn with understanding. It is to be remembered that the head teacher is the only person who has easy access to all classrooms and who will be aware of the kinds of activity going on in each.

The lack of confidence amongst primary teachers in teaching mathematics has been referred to several times (e.g. Ward, 1979; Ray, 1975) and the lack of balance in the mathematics curriculum (as well as other areas) caused H.M.I.'s to suggest that some rather idiosyncratic decision-making was taking place in primary schools (Great Britain, D.E.S., 1978a). This suggests that head teachers may not be

adequately aware of the importance of such factors as regularity of staff meetings, for example, and may too readily
assume that the norms selected by them to determine the ethos
of the school (Blyth, 1965) are clear to all members of staff
when in fact they are not. It may be that regular, formal
staff meetings are not the answer but it would seem that a
fair proportion of primary school head teachers need further
guidance than they already enjoy, in providing leadership
within a school. This may be particularly so with respect
to developing further awareness of the repercussions of their
attitudes to, and decisions upon, curricular matters.

If a head teacher has the opportunity to appoint a
mathematics co-ordinator, then it is part of the head's
responsibility as leader of the school staff to ensure that
such a person receives relevant in-service training to
satisfy the demands of that job, and fully to support the
mathematics co-ordinator within the school. Again, the fact
that work done has been judged to be noticeably effective in
only a quarter of such posts that exist (Great Britain,
D.E.S., 1978a) suggests that either the training given (if
any) was not effective, or that there was inadequate support
from the head teacher within the school, or a combination of
both these factors. It would seem that in some instances
where such courses are given, they appear to result in little
and slow improvement (Cockcroft submission, B18). If there
is not provision for a mathematics co-ordinator on a school
staff then again the responsibility lies with the head
teacher to give a lead in ensuring that a balanced mathematics curriculum is implemented, and to support staff fully
in doing so.

The head of the mathematics department provides curricular leadership at secondary level and evidence suggests that
the degree of effectiveness of mathematics teaching in
secondary schools is directly related to the quality of the
head of department (Neill, 1978; Cockcroft submissions, B25,
J61). This is a crucial role and it would appear that it is
a post which brings with it feelings of anxiety, futility and
the mistrust of fellow staff members, for which the guidance
provided is inadequate (Hall and Thomas, 1977).

It would clearly be advantageous for the heads of
mathematics departments to receive special in-service training
that would help them to identify problems that are specific
to their subject department and the effective running of it.
The proportion of non-specialist teachers of mathematics in
secondary schools in itself presents a special problem and
may add to the difficulty of drawing the department together
as a working unit. With adverse attitudes to mathematics
generally on the part of those teaching it, as well as the
taught, it could well be important for the department head to
involve members of the department in more curricular planning
in order to help create a stronger feeling of unity and

identification. However, there appears to be some evidence
of reluctance on the part of heads of department to hold
regular meetings which could lead to such unity (Hall and
Thomas, 1977) and hence, it would seem, possibly also a
reluctance to encourage such involvement and to delegate
responsibility. As well as concern for the departmental unit,
heads of department are responsible for probationary teachers
and for their introduction into the 'subject culture'
identified by Hoad (1974) within a school.

Support Roles

Other teachers can not only control and lead, they can
also act as *support*. In secondary schools, the notion of the
department as the unit seems to offer much promise. A good
departmental team clearly can take much of the pressure off
individual teachers, particularly in helping probationers
make it through the difficult first stages. The department
also seems to be the more appropriate mediator of outside-
school influences than is the individual teacher. Corporate
decision-making through regular meetings about curricular
priorities and emphases can release more 'thinking-space' for
the individual teacher to handle micro-curricular, and other
pedagogical decisions in their own classrooms. Again, much
depends on the quality of the head of department not just as
a leader of the support team but also as the representative
of the mathematics department in negotiations with the rest
of the school's upper hierarchy.

The department can play a further role in alleviating
another source of a mathematics teacher's stress which is
conflict with teachers of 'user' subjects. The D.E.S.
(Great Britain, 1979b) report recommends that 78 per cent of
all schools need to foster closer links with other subject
departments within the school. If there is such a deficiency
with respect to this kind of co-operation, it is one that
could at least partially be rectified by school-based in-
service training that brings mathematics departments together
with staff of other departments. There is a difficulty for
mathematics teachers in that, on the one hand, they are being
asked to identify themselves more strongly with the mathematical community (Fletcher, 1975) but, at the same time, to
become more aware of the mathematical needs of other subject
areas. These should not be seen as conflicting demands since
an increase in their sensitivity to the relevance of their
subject to other subjects can only add to their professionalism as mathematics teachers (Otte, 1979). One way that
such links could be forged is through curriculum development
within schools which involves a variety of disciplines, for
example, physical or social sciences. Teachers of subjects
which use mathematics are also by definition teachers of
mathematics. The perceived relevance of the subject by
pupils as a result of such co-operation could only benefit
the mathematics teachers, not to mention the pupils.

The need for support in the primary school is equally great due largely to the prevalence of feelings of insecurity experienced by many primary teachers with respect to the teaching of mathematics (Great Britain, D.E.S., 1978a; Straker, 1978; Ward, 1979). Ray's (1975) study indicating the favourable attitudes towards mathematics of most of his sample of college of education entrants, but poor attitudes towards the *teaching* of it arising from the dislike of methods by which they themselves had been taught, again raises the spectre of the vicious circle which exists within mathematics education. The roles of the primary head teacher and the teacher with the post of responsibility for mathematics in the primary school have already been discussed. It would seem that the size of the problem is such, however, that they will need considerable help from outside the school in the form of advisory staff. There may be a lesson to be learned from American experience here, which suggests that where teachers themselves identify their problems and are given consistent help, in school, over a period of time, success in building up their confidence ensues (Easley, 1980). It is essentially a matter of the advisory person gaining the trust of the classroom teacher, which cannot happen in single visits, widely spaced in time.

The Physical conditions

A major constraint, within secondary schools, concerns the actual physical provision for the teaching of mathematics. There is great irony in the fact that mathematics is recognized as a high status subject (Gordon, 1978; Hall and Thomas, 1977) yet, in many schools, there is poor provision for the teaching of it in terms of specialist rooms and resources. The D.E.S. (Great Britain, 1980a) recommends the re-allocation of accommodation in 27 per cent of all schools and, possibly more disturbing, in 45 per cent of grammar schools. Mention is made in the report of the desirability of mathematics teachers coming to view their subject (and hence the teaching of it) from a more open and creative perspective. It is futile to hope for such a change of attitude without also changing the circumstances in which many mathematics teachers work. Specially allocated rooms are important to the identification of mathematics as a subject, the teaching of which requires more than just chairs, desks and a blackboard. It has been noted in one instance at least that the 'nomadic existence' of some mathematics departments has led to a restrictive attitude on the part of teachers towards their lessons and a lack of concern for the display of pupils' work, which may suggest that little value is ascribed to it (Cockcroft submission, J51).

Mixed ability teaching, to be successful, requires a good variety of resources (Lingard, 1976). The D.E.S. (Great Britain, 1979b, 1980a) does not call for more resources as such but, rather, greater use of those already existing in about 30 per cent of all secondary schools. It may be, however, that these are not being used because they are not centrally available in an area designated for the teaching of mathematics specifically.

There is a strong case for locating mathematics departments in a specialist area within schools in order to bring about more effective teaching of the subject. An important consequence as noted above, would be the way in which this could help favourably to alter the perceptions of both teachers and pupils of mathematics and hence the teaching and learning of it. However, space allocation is not usually the province of the individual mathematics teacher, and it is perhaps worth pointing out that it is unlikely that those who do have control over such matters are mathematically trained. Much therefore depends on the awareness of heads, deputy heads, governors and advisers about the need for specialist accommodation, and also on the political skill and resourcefulness of heads of departments to negotiate successfully within their schools.

Summary

Study of the social context of mathematics education therefore makes us aware initially of the roles of those, *other than the mathematics teacher,* who clearly affect the quality of mathematics teaching. Lortie (1975) neatly sums up the individual teacher's frustration: 'Teachers have a built-in resistance to change because they believe that their work environment has never permitted them to show what they can really do. Many proposals for change strike them as frivolous ...' (p.235). There is clearly an assumption behind such proposals that the 'fault' lies with the individual teacher. Lortie's teachers clearly feel that the 'fault' lies with the conditions which surround and limit them. The research which we have reviewed says a great deal in support of Lortie's case.

What now can the research tell us about the individual teachers and their classroom work?

INTERNAL CONSTRAINTS: THE INDIVIDUAL TEACHER

Perceptions of content

There is much public concern expressed about teachers' and prospective teachers', knowledge of mathematics but research into the social context makes plain the need to focus more on the *attitudes* and *perceptions* of teachers with respect to the mathematical content of the curriculum.

It is commonplace that the hierarchical nature of mathematics can easily impose a rigid structure on the way in which the subject is taught. It may too easily be accepted as 'given' and tend to constrain teachers to present content in a particular order as well as in a particular manner. This may be what is seen to convey the 'academic' nature of the subject to which Richardson (1975) refers and is, no doubt, part of the reason the subject has high status ascribed to it. But content has a social meaning also and it is this view which allows the pupils really to get to grips with mathematics. As Bauersfeld (1980) points out 'Teaching and learning mathematics is realized through *human interaction*' (p.35, author's italics) and teachers need to remember that the mathematics classroom, like any other, is a place for dialogue with and between pupils. Mathematical meaning can be negotiated with and for them, just as the whys and wherefores of the Battle of Hastings can be, and if this happens the atmosphere becomes more one of inquiry and discussion. This is a difficult task, but it seems important that mathematics teachers recognize that, as within the curriculum generally, different kinds of tasks, related to different kinds of mathematical content, require different kinds of teacher control in the classroom (Bossert, 1979).

For example, at primary level, a mechanical view of the nature of mathematics is likely to result in the teacher acting as a purveyor of mathematical facts, with pupils performing repetitive tasks in a somewhat passive manner. On the other hand, primary teachers who are aware of, and able to identify, the processes inherent in the formation of mathematical concepts are likely to approach the teaching of the subject in quite a different way. This perception of the nature of mathematics clearly will result in a more varied classroom atmosphere, characterized in some degree by activity and inquiry.

Perceptions of pupils

Closely related to attitudes to content are the teachers' perceptions, and indeed constructions, of the pupils in their care. If one had to choose *the* most significant controlling variable emerging from the research surveyed it would be this one. Time and again we read of the powerful influence of teachers' views of their pupils. As Roberts (1971) says 'The teacher must be made aware of the potency of his expectations. Research shows that very, very simple acts on the part of teachers result in astonishing behavioural changes in students' (p.174).

These perceptions and expectations are most significant in a 'visible' subject like mathematics where success and failure is all too obvious. The particular pupils at risk appear to be girls and the generally less able children, although all pupils can be affected. The problems also seem to be greater at secondary level where the teachers have less opportunity for contact with their charges than at primary level. Judgments are therefore made on the basis of minimal evidence (Hargreaves, 1967) and, by means of 'self-fulfilling prophecy', can clearly inhibit the pupils' mathematical progress.

The institutional setting of the classroom which changes 'children' into 'pupils' faces them with a situation in which they have to learn to become a member of a competitive group (Jackson, 1968), and to be identified with the 'community of knowledge' in the class with respect to the relative ability of each member of it (Nash, 1973). Success or failure in learning mathematics is an obvious criterion for the judgment of pupils by teachers and peers so that the "perceived self" as seen by others (Hudson, 1968) is made explicit to the pupils themselves. Overt judgments of this nature and the more obvious competitive aspect of the subject may often have the effect of causing them to 'switch off' where mathematics is concerned, having too readily identified themselves as failures. Ray's (1975) study of the attitudes of student teachers to their own school experiences is a good example. His results noted particularly the case of girls whose attitudes were characterized with references to 'humiliation' and 'not a girl's subject'. It is not, however, always failure that causes such an effect, for there is evidence that girls may not wish to be seen to succeed in mathematics because of the supposed masculine overtones the subject has (Horner, 1968).

Another aspect of teachers' perceptions of their pupils concerns their potential for mathematics. It may be, as Selkirk (1974) postulates, that pupils have a quite definite 'ceiling' perhaps related to the levels of abstraction which they meet in the upper levels of the subject and beyond which they possibly should not realistically be forced to go. It is more likely that the teachers' perceptions set the 'ceiling' for the pupils. At its crudest, for example, the poorly spoken pupil may be dismissed as having little mathematical ability while the articulate pupil may be assumed to be mathematically able when the situation may in fact be the reverse. In either case, the individual's mathematical learning needs will not be identified or satisfied. Thus, if the vicious circle is indeed closed, the pupils will perform according to the teacher's expectation because of what Brophy and Good (1974) call 'the credibility of the source' of the judgement made upon them.

Teaching individuals

Schooling is essentially a compromise between a possible ideal of individual tuition for every child and the availability of resources, both financial and human. The result is a typical class of 20-30 pupils all studying roughly the same material. Regardless of who makes the macro-curricular choices of content, the micro-curricular adaptations are in the hands of the individual teacher. In the primary school, with considerable contact time, the teacher does have a chance to see the richness of each individual child. It is possible therefore for a primary teacher to make many of the adaptations in curriculum and methodology necessary to extend each child. The fact that this may not happen can be due to many reasons - the teacher's limited perception and knowledge of mathematics, the teacher's inability to judge the child's mathematical potential, the teacher's lack of awareness of a rich pedagogical repertoire etc. This appears to be the case particularly with more able pupils in primary schools (Great Britain, D.E.S., 1978a).

The problem is different, but much more acute in secondary schools. Because of the reduced contact time for teaching, most secondary school teachers have only a shallow knowledge and limited perception of their pupils and it is quite likely that many mathematics teachers see their differences in mathematical ability in terms of their *rate* of learning. This is the reason for much reference to slower pupils and to pupils who 'catch on quickly'. If the teaching must be completed in 40 minutes (say) then children will distribute themselves into three groups, those who finish early, those who just finish and those who never finish in time. If that is the perception teachers have of pupils, they will not be provoked into developing a richer pedagogical repertoire, and the vicious circle continues. Limited perceptions reinforce limited methods which in turn fulfill the limited expectations.

The advent of mixed ability teaching has provoked two main responses, either grouping pupils by ability within classes, or a move to greater individualization. The latter development relies heavily on the availability of prepared materials, which being perceived by the teachers as 'self-explanatory' can have the undesirable effect of coming between the teacher and the pupil, and eliminating the need for explanatory dialogue (Morgan, 1977). This dilemma may not be the only reason that the mathematical needs of the less-able are not being met (Great Britain, D.E.S., 1979b) but it is likely to be a major contributing factor. It is clear from the strength of the recommendation made by the D.E.S. (Great Britain, D.E.S. 1980a) for new courses to be organized for less able pupils in 68 per cent of comprehensive schools, that the teaching methods adopted are not achieving an acceptable degree of success. Whether more resources will solve this problem is debatable. The real solution, as has already been mentioned, lies in *increasing* the amount of teacher pupil contact so that the teacher is encouraged to construct a richer 'picture' of that child, and in the case of the less able child, to be able to identify hidden strengths and abilities as well as the more obvious weaknesses and disabilities. If more resources can create more contact time, then the problem would become manageable. If resources come between the teacher and the pupil, the problem becomes more intractable.

Teacher stress

That teaching is a stressful activity is undeniable and many studies illustrate the problems of the moment-by-moment decision-making required of teachers (Hargreaves, 1972; Jackson, 1968). The effect of mathematics as a subject has already been referred to in connection with how teachers view mathematical content, and also the effect of their perceptions upon pupils. The 'visibility' identified in connection with it can add to the normal stresses a teacher would expect within a classroom. Just as success or failure is evident on the part of pupils learning mathematics, so is the success or failure on the part of teachers to teach it. Pupils are quick to sense when it is not just a small number of them who do not understand what is being taught. They therefore may turn on the teacher and the 'bewildered anarchy' or 'corporate hostility' to which Blyth (1965) refers can quickly build up. Clearly this happens with teachers of other subjects as well but it seems appropriate to draw attention to the fact that it is likely to happen in mathematics lessons more often than most, because of the 'public' aspect of the criteria of successful or unsuccessful teaching (made particularly obvious where the teacher is not confident in teaching the subject). As Hargreaves (1972) suggests, discipline and instruction become inseparable in a classroom and where instruction fails, undisciplined behaviour will follow.

Research into teacher stress suggests that pressure of time is one of the principal components leading to stress (Kyriacou and Sutcliffe, 1978), and clearly in the secondary mathematics classroom, with limited time available to help individual pupils, teacher frustration can be great. A further source of stress is located in the teacher's modelling behaviour (Good and Brophy, 1978), which refers to the teacher's behaviours which act as models for the pupils. They can concern problem-solving behaviours or personality behaviours - in fact anything the teacher does is there as a potential model to be copied by a pupil. Mathematics teachers who are hypercritical of poor pupil performance can produce a destructive classroom climate because, Good and Brophy point out, 'The students imitate such teachers, even though they dislike them, because the teacher not only models but also rewards such behaviour' (p.123). Also if the teacher is lacking mathematical confidence this will show in their behaviour and can easily be adopted by the pupils. Other problems can surround the teacher's credibility with his pupils. As Good and Brophy (1978) explain, 'Teachers may not only have to model appropriately by practising what they preach, they may have to call the students' attention to their own credibility' (p.135). Ensuring that one's behaviour matches one's statements, appearing as a good model of

66

'mathematical behaviour', and appearing as a tolerant and fair judge of performance are difficult skills to monitor and practise at the best of times. In the public arena of the classroom, with all the other pressures of time and curriculum coverage present, they clearly represent a potentially great source of stress for every teacher.

OUTSTANDING PROBLEMS

In the second half of this chapter we have attempted to summarize our conclusions from the reviewed research which focuses on the teacher. From a 'social' perspective, the major problems facing primary teachers seem to differ from those facing secondary teachers.

In the case of primary teachers of mathematics, the problems appear to centre on the teachers' lack of knowledge of, poor attitude to, and limited perception of, the mathematics curriculum. These deficiencies manifest themselves in many aspects, but two areas seem worth emphasizing, (1) the complexity of curricular decisions which primary teachers must make, in the absence of adequate guidance from head teachers, specialist mathematics co-ordinators, or mathematics advisers, and (2) the problems of extending the more able pupils.

In the secondary school the major problems seem to revolve around the shallowness of the teachers' perceptions of their pupils. Two corollaries of this are, (1) a tendency to create unfavourable attitudes towards the learning of the subject, and (2) the generally unsuccessful teaching of less able pupils through what might be called methodological simplicity.

CONCLUDING REMARKS

The amount of research on the social context of mathematics education is extremely limited. Many of our conclusions are based upon surveys, analyses and extrapolations from results of research not carried out with specific reference to mathematics. Nevertheless, these studies have sensitized us to the significance of several factors within the social context of schools which exert a powerful influence on the quality of mathematical learning. Our overriding conclusion, therefore, is that mathematics education research should be directed away from the individual child as a learner and towards an increased understanding of the effects of the social context of schools on the learning of mathematics. Paradoxically, in doing so it is likely that greater insight could be gained into the causes of the difficulties faced by the individual child learning mathematics.

References

AINSWORTH, M.E. and BATTEN, E.J. (1974): *The effects of environmental factors on Secondary Educational attainment in Manchester: a Plowden follow-up.* London: Macmillan Education Ltd.

AMIDON, E.J. and HOUGH, J.B. (Eds) (1967): *Interaction Analysis: Theory, Research and Application.* Reading (Mass.): Addison-Wesley Publishing Co.

APPLE, M.W. (1980): 'The other side of the hidden curriculum: Correspondence theories and the labor process', *Journal of Education,* 162, 47-66.

ARMSTRONG, J.M. (1980): *Achievement and Participation of Women in Mathematics.* Report of a two-year study funded by the National Institute of Education (Report 10-MA-OO), Denver, Colorado, Educational Commission of the States.

ARFWEDSON, G. (1976): 'Ideals and Reality of Schooling', *Schriftenreihe des IDM,* Universität Bielefeld, 6, 139-146.

ASHEROOK, A. (1977): 'Teaching Mathematics to Gifted Children', *Trends in Education,* 2, 9-13.

ASHTON, P., KNEEN, P., DAVIES, F. and HOLLEY, B.J. (1975): *The aims of primary education: a study of teachers' opinions.* London: Macmillan Education.

AUSTIN, J.L. and HOWSON, A.G. (1979): 'Language and Mathematical Education', *Educational Studies in Mathematics,* 10, 161-97.

BARNES, D. (1971): 'Language and Learning in the Classroom', *Journal of Curriculum Studies,* 3, 1, 29-38.

BAUERSFELD, H. (1980): 'Hidden Dimensions in the So-called Reality of a Mathematics Classroom', *Educational Studies in Mathematics,* 11, 1, 23-41.

BEGLE, E.G. (1979): *Critical Variables in Mathematics Education.* Washington: Mathematical Association of America and the National Council of Teachers of Mathematics.

BELSOM, C.G.H. and ELTON, L.R.B. (1974): 'The effect of syllabus in mathematical knowledge', *Physics Education,* 9, 462-3.

BENNET, N. (1976): *Teaching Styles and Pupil Progress.* London: Open Books.

BERNSTEIN, B. (1971): *Class, Codes and Control.* London: Routledge and Kegan Paul, Vol. 1.

BERNSTEIN, B. (1975): 'Class and Pedagogies' Visible and Invisible', *Educational Studies,* 1, March, 23-41

BERRILL, R., and WALLIS, P. (1976): 'Sex roles in Mathematics', *Mathematics in Schools,* March, p.28.

BISHOP, A.J. and McINTYRE, D.I. (1969): 'A comparison of secondary and primary teachers' opinions regarding the content of primary school mathematics', *Primary Mathematics,* 7, 2, August, 33-9.

BISHOP, A.J. and McINTYRE, D.I. (1970): 'A comparison of teachers' and employers' opinions regarding the content of secondary school mathematics', *Mathematical Gazette,* 54, 389, 229-33.

BISHOP, A.J. (1975): *Opportunities for attitude development within lessons.* Paper presented at I.C.M.E. conference on 'Attitudes towards Mathematics', Hungary, September.

BLYTH, W.A.L. (1965): *English Primary Education: A sociological description Vol. 1.* London: Routledge and Kegan Paul.

BOSSERT, S.T. (1979): *Tasks and Social Relationships in Classrooms.* Cambridge: Cambridge University Press.

BROOKES, W.M. (1978): 'Interpretation. the Hermeneutic approach'. In: YATES, J. *Four Mathematical Classrooms: An enquiry into Teaching Methods.* University of Southampton, vi-xiii.

BROPHY, J.E. and GOOD, T.L. (1974): *Teacher-Student Relationships: Causes and Consequences.* New York: Holt, Rinehart and Winston.

BROWN, S. and McINTYRE, D. (1978): 'Factors influencing teachers' responses to curricular innovations'. *Research Intelligence,* 4, 1, 19-24.

BURNS, R. (1976): 'Preferred teaching approach in relation to self and other attitudes', *Durham Research Review,* 7, Spring.

BURTON, G.M. (1978): 'Mathematical ability - is it a masculine trait?', *School Science and Mathematics,* 78, 566-74.

CLARK, C.M. and YINGER, R.J. (1980): *The Hidden World of Teaching: Implications of Research on Teacher Planning.* Paper presented to the American Educational Research Association, Boston.

CLIFT, P.S., CYSTER, R., RUSSELL, J. and SEXTON, B. (1978): 'The use of Kelly's Repertory Grid to Conceptualize Classroom Life'. In: McALEESE, R. and HAMILTON, D. (Eds) *Understanding Classroom Life.*

COHEN, L. (1976): *Educational Research in Classrooms and Schools.* London: Harper and Row.

COOPER, D. and EBBUTT, D. (1974): 'Participation in action research as an in-service experience', *Cambridge Journal of Education,* 4, 2, 65-71.

CORNELIUS, M.L. (1973): 'The new graduate mathematics teacher in school', *Mathematical Education for Teaching,* 1, 2, 10-15.

COX, T. (1979):'A follow-up study of reading attainment in a
          sample of 11 year old disadvantaged children',
          Educational Studies, 5, 1, 53-60.
DAWE, L.C.S. (1978): 'Teaching Mathematics in a Multicultural
          School', *Forum of Education,*37, 2, June, 24-31.
DELANEY, K. (1977): 'Suppose that', *A.T.M. Supplement 20,*
          October, pp.2-3.

DOCKRELL, W.B. and HAMILTON, D. (Eds) (1980): *Rethinking
          Educational Research.* London: Hodder and
          Stoughton.
DRAPER, A.C. (1974): 'The professional education of mathe-
          matics teachers in main mathematics courses in
          colleges of education',*Mathematical Education for
          Teaching,* 3, 10-16.
DOUGLAS, J.W.B. (1967): *Home and the School.* London:
          MacGibbon and Kee.
DUCKWORTH, D. and ENTWISTLE, N.J. (1974): 'Attitudes to
          School Subjects: A repertory grid technique',
          British Journal of Educational Psychology,
          44, 1, 76-82.
DUDLEY, B.A.C. (1975): 'Bringing Mathematics to Life',
          Journal of Biological Education, 9, 6, 263-8.
EASLEY, J.A. Jnr. (1975): 'Thoughts on individualized
          instruction in mathematics', *Schriftenreihe des
          IDM,* Universität Bielefeld, No. 5, 21-48.
EASLEY, J.A. (1980): *School options for the use of a team of
          resource persons (RPS) in primary grade mathe-
          matics teaching.* University of Illinois, Urbana.
EDELMAN, M. (1974): 'The Political Language of the Helping
          Professions', *Politics and Society,* 4, Fall, 1974,
          295-310.
EGGLESTON, J. (Ed) (1974): *Contemporary Research in the
          Sociology of Education.* London: Methuen.
EGGLESTON, J. (Ed) (1979): *Teacher Decision-making in the
          Classroom.* London: Routledge and Kegan Paul.
ELLIOTT, J. and ADELMAN, C. (1975): 'Teachers' Accounts and
          the Control of Classroom Research', *London
          Educational Review,* Autumn, 29-37.
FENNEMA, E. (1979): 'Women and Girls in Mathematics - Equity
          in Mathematics Education', *Educational Studies in
          Mathematics,* 10, 4, 389-401.
FIELKER, D.S. (1979): Editorial in *Mathematics Teaching,*
          No. 86, March, pp.2-3.
FINLAYSON, D. and QUIRK, S. (1979): 'Ideology, reality
          assumptions and teachers' classroom decision-
          making'. In: EGGLESTON, J. (Ed). *Teacher
          Decision-making in the Classroom.* London:
          Routledge and Kegan Paul, pp. 50-73.
FITZGERALD, A. (1978): 'Corridor of Power', *Mathematics in
          School,* 7, 1, 23-25.
FLANDERS, W.A. (1967): 'Some relationships among teacher
          influence, pupil attitudes and achievement'. In:
          AMIDON, E.J. and HOUGH, J.B. (Eds) *Interaction*

Analysis: Theory, Research and Application.
Reading (Mass.): Addison-Wesley.

FLETCHER, T.J. (1975): 'Is the teacher of mathematics a
mathematician or not?', *Schriftenreihe des IDM*,
Universität Bielefeld, No. 6, pp. 203-18.

FRANCIS, P. (1975): *Beyond Control? A study of discipline in
the comprehensive school*. London: George Allen
and Unwin.

FREEMAN, D.J. and KUHS, T.M. (1980): *The Fourth Grade
Mathematics Curriculum as Inferred from
Textbooks and Tests*. Paper presented to the
annual meeting of the American Educational
Research Association, April 7-11, Boston.

GOOD, T.L. and BROPHY, J.E. (1978): *Looking in Classrooms*.
New York: Harper and Row.

GORDON, P. (1978): 'Control of the Curriculum'. In: LAWTON
et al. (Eds) *Theory and Practice of Curriculum
Studies*. London: Routledge and Kegan Paul.

GRACE, G. (1978): *Teachers, Ideology and Control: A Study in
Urban Education*. London: Routledge and Kegan
Paul.

GRAY, J. and SATTERLY, D. (1976): 'A chapter of errors:
Teaching styles and pupil progress in retrospect',
Educational Research, 19, 2, 45-56.

GREAT BRITAIN. DEPARTMENT OF EDUCATION AND SCIENCE (1967):
Children and their Primary Schools (The Plowden
Report). London: H.M.S.O.

GREAT BRITAIN. DEPARTMENT OF EDUCATION AND SCIENCE (1972):
Teacher Education and Training (The James Report).
London: H.M.S.O.

GREAT BRITAIN. DEPARTMENT OF EDUCATION AND SCIENCE (1975):
A Language for Life (The Bullock Report). London:
H.M.S.O.

GREAT BRITAIN. DEPARTMENT OF EDUCATION AND SCIENCE (1978a):
*Primary Education in England: A survey by
H.M. Inspectors of Schools*. London: H.M.S.O.

GREAT BRITAIN. DEPARTMENT OF EDUCATION AND SCIENCE (1978b):
*In-service education and training for teachers:
A basis for discussion*. London: H.M.S.O.

GREAT BRITAIN. DEPARTMENT OF EDUCATION AND SCIENCE (1979a):
Mathematics 5-11: A handbook of suggestions.
London: H.M.S.O.

GREAT BRITAIN. DEPARTMENT OF EDUCATION AND SCIENCE (1979b):
Aspects of secondary education in England.
London: H.M.S.O.

GREAT BRITAIN. DEPARTMENT OF EDUCATION AND SCIENCE (1980a):
*Aspects of secondary education in England:
Supplementary information on Mathematics*.
London: H.M.S.O.

GREAT BRITAIN. DEPARTMENT OF EDUCATION AND SCIENCE (1980b):
*Mathematical Development: Primary Survey Report
No. 1*, (Assessment of Performance Unit).
London: H.M.S.O.

GRIFFITHS, H.B. and HOWSON, A.G. (1974): *Mathematics: Society and Curricula*. Cambridge: Cambridge University Press.

HALL, J.C. and THOMAS, J.B. (1977): 'Research Report: Mathematics Department Headship in Secondary Schools', *Educational Administration*, 5, 2, 30-7.

HALL, J.C. and THOMAS, J.B. (1978): 'Role Specification for Applicants for Heads of Mathematics Departments in Schools', *Educational Review*, 30, 1, 35-9.

HARGREAVES, D. (1967): *Social Relations in a Secondary School*. London: Routledge and Kegan Paul.

HARGREAVES, D. (1972): *Interpersonal Relations and Education*. London: Routledge and Kegan Paul.

HEIMER, R.T. and NEWMAN, S. (1965): *The New Mathematics for Parents*. New York: Holt, Rinehart and Winston.

HILSUM, S. and START, K.B. (1974): *Promotion and Careers in Teaching*. Slough: NFER.

HILSUM, S. and STRONG, C. (1978): *The Secondary Teacher's Day*. Windsor: NFER.

HOAD, P. (1974): A study of the professional socialisation of graduate student teachers. Unpublished dissertation for the degree of Ph.D., University of Sussex.

HORNER, M.S. (1968): Sex differences in achievement motivation and performance in competitive and non-competitive situations. Unpublished doctoral dissertation, University of Michigan.

HOWSON, A.G. (1975): 'Teacher involvement in curriculum development', *Schriftenreihe des IDM*, 6, pp.267-87.

HOWSON, A.G. (1980): 'Socialist mathematics: does it exist?' *Educational Studies in Mathematics*, 11,3, 285-99.

HOYLE, E. (1979): Research on the Professional Development of Teachers. Unpublished paper.

HUDSON, L. (1968): *Frames of Mind*. London: Methuen.

JACKSON, P.W. (1968): *Life in Classrooms*. New York: Holt, Rinehart and Winston.

JUDGE, H. (1975): 'How are we to get better teachers?', *Higher Education Review*, 8, 1, 3-16.

KALLOS, D. and LUNDGREN, U.P. (1976): 'Purpose and Comments', *Materialien und Studien Band 6*, IDM Universität Bielefeld, 3-12.

KEDDIE, N. (1971): 'Classroom knowledge'. In: Young M.F.D. (Ed) *Knowledge and Control*. London: Collier-Macmillan.

KERR, E. (1977): 'Some thoughts on the educational system and mathematics teaching', *The Mathematical Gazette*, 61, 157-73.

KERRY, T. (1978): 'Bright Pupils in Mixed Ability Classes', *Research Intelligence*, 4, 2, 103-14.

KYRIACOU, C. and SUTCLIFFE, J. (1978): Teacher stress: prevalence, sources and symptoms', *British Journal of Educational Psychology*, 48, 159-67.

73

LEZOTTE, L.W. and PASSALACQUA, J. (1978): Individual School
        Buildings do Account for Differences in Measured
        Pupil Performance. Occasional Paper No. 6,
        Institute for Research on Teaching, College of
        Education, Michigan State University.
LINGARD, D.W. (1976): 'Teaching Mathematics in Mixed Ability
        Groups'. In: WRAGG, E.C. (Ed) *Teaching Mixed
        Ability Groups*. London: David and Charles.
LORTIE, D. (1975): *Schoolteacher*. Chicago: The University
        of Chicago Press.
LUCHINS, E.H. (1979): 'Sex differences in mathematics: How
        not to deal with them', *American Mathematical
        Monthly,* 86, 161-8.
LUMB, D. (1974): 'Student teachers and mathematics:
        Mathematical Competence', *Mathematics Teaching,*
        No. 68, 48-50.
LUMB, D. and CHILD, D. (1976): 'Changing Attitudes to the
        Subject, and the Teaching of Mathematics amongst
        Student Teachers', *Educational Studies,* 2, 1-10.
MACLURE, S. (1968): *Curriculum innovation in practice.*
        London: H.M.S.O.
MARCUS, A.C. *et al.* (1976): Administrative leadership in a
        sample of successful schools. Unpublished Paper
        AERA meeting, April, San Francisco. ERIC ED 125123.
MELLIN-OLSEN, S. (1976): Instrumentalism as an Educational
        Concept. Det Pedagogiske Seminar 1976,
        Universitetet Bergen, Norway.
MIES, T., OTTE, M., REISZ, V., STEINBRING, H. and VOGEL, D.
        (1975): 'Tendencies and Problems of the training
        of Mathematics Teachers', *Materials des IDM,*
        Universität Bielefeld, 6.
MORGAN, J. (1977): Affective Consequences for the Learning
        and Teaching of Mathematics of an Individualised
        Learning Programme. DIME Projects, Department
        of Education, University of Stirling.
MORRISON, A. and McINTYRE, D. (1969): *Teachers and Teaching.*
        Harmondsworth: Penguin Books.
MORRISON, A., McINTYRE, D. and SUTHERLAND, J. (1965):
        'Teachers Personality Ratings of Pupils in
        Scottish Primary Schools', *British Journal of
        Educational Psychology,* 35, 3, 306-19.
MORRISON, A., McINTYRE, D. and SUTHERLAND, J. (1966):
        'Social and educational variables relating to
        teachers' assessment of primary school pupils',
        British Journal of Educational Psychology,
        36, 3, 272-9.
MUSGRAVE, P.W. (1979): *The Sociology of Education,* third
        edition. London: Methuen.
MUSGROVE, F. and TAYLOR, P.H. (1969): *Society and the
        Teacher's Role.* London: Routledge and Kegan Paul.
McALEESE, R. and HAMILTON, D. (Eds) (1978): *Understanding
        Classroom Life.* Windsor: NFER.
McINTOSH, A. (1979): 'When will they ever learn?', *Mathe-
        matics Teaching,* 19, No. 86, March, i-iv.

McKEACHIE, W.J., POLLIE, D. and SPEISMAN, J. (1955): 'Relieving Anxiety in Classroom Examinations', *Journal of Abnormal and Social Psychology,* 93-8.

NASH, R. (1972): 'Measuring teacher attitudes', *Educational Research,* 14, 141-6.

NASH, R. (1973): *Classrooms Observed.* London: Routledge and Kegan Paul.

NASH, R. (1974): 'Pupils' Expectations of Teachers', *Research in Education,* No. 12, November, 47-61.

NEILL, H. (1978): Report on Durham Mathematics Appointment. Unpublished paper.

NEWBOLD, D. (1977): *Ability Grouping - the Banbury Enquiry.* Windsor: NFER.

NISBET, J.W. (1979): 'Schools and Industry - specialised in-service training for teachers', *Trends in Education.* Summer Issue 2, 4-7.

NISBET, J.W. (1980): 'Educational Research: The State of the Art'. In: DOCKRELL, W.B. and HAMILTON, D. (Eds) *Rethinking Educational Research,* p.5. London: Hodder and Stoughton.

OTTE, M. (1979): 'The education and professional life of mathematics teachers', *New Trends in Mathematics Teaching,* 4, 107-33. UNESCO.

POPHAM, W.J. (1971): 'Performance tests of Teaching Proficiency: Rationale, Development and Validation', *American Educational Research Journal,* 8, 105-17.

PREECE, M. (1979): 'Mathematics; the Unpredictability of Girls?', *Mathematics Teaching,* No. 87, June, 27-9.

PREECE, P.F.W. (1977): 'Problems of discipline on teaching practice: a model based on catastrophe theory', *Research Intelligence,* 3, 2, 22-3.

RAY, S.P. (1975): Some factors affecting recruitment to main mathematics courses in a College of Education. Unpublished M.Ed. thesis, University of Newcastle upon Tyne.

REID, W.A. (1979): Making the Problem fit the Method: a Review of the 'Banbury Enquiry', *Journal of Curriculum Studies,* 11, 2, 167-73.

RESEARCH GROUP OF ASSOCIATION OF TEACHERS OF MATHEMATICS (1973): *Focus on teaching.* A.T.M., Nelson, Lancs.

REVUZ, A. (1978): 'Changes in the teaching of Mathematics in France'. *Educational Studies in Mathematics,* 9, 171-81.

RICHARDSON, E. (1975): *Authority and Organization in the Secondary School.* London: Macmillan Educational.

ROBERTS, J.I. (1971): *Scene of the Battle: Group Behavior in Urban Classrooms.* New York: Doubleday.

ROSENSHINE, B. (1971): *Teaching Behaviours and Student Achievement.* Windsor: NFER.

ROSENTHAL, R. and JACOBSON, L. (1968): *Pygmalion in the Classroom: Teacher expectation and pupils' intellectual development.* New York: Holt, Rinehart and Winston.

THE ROYAL SOCIETY (1976): *The Training and Professional Life of Teachers of Mathematics.* The Royal Society, November.

RUTTER, M., MAUGHAN. J., MORTIMORE, P. and OUSTON, J. (1979): *Fifteen Thousand Hours: Secondary Schools and their effects on children.* London: Open Books.

SCHOOLS COUNCIL (1967): Schools Council Working Paper 10 Curriculum Development: Teachers' Groups and Centres.

SCHOOLS COUNCIL (1977): *Mixed-ability teaching in mathematics.* London: Evans/Methuen Educational.

SELKIRK, J. (1974): Academic aspects of pupils' choice of Advanced level studies, with particular reference to specialisation and the choice of mathematics. Unpublished Ph.D. dissertation. University of Newcastle-upon-Tyne, D11212/75.

SHUARD, H.B. (1973): 'A pilot survey of the expectations of Heads of Mathematics Departments about new Mathematics Teachers', *Mathematical Education for Teaching,* 1, 2, 16-20.

SHUARD, H.B. (1975): 'Old and New-style courses in Mathematics. *Mathematical Education for Teaching',* 2, 1, 13-18.

SHUARD, H.B. (1977): 'A survey of the present state of teacher education in Mathematics', *Mathematical Education for Teaching,* 2, 4, 28-32.

SHUARD, H.B. (1978): 'The 1977 Survey of Teacher Education Courses in Mathematics', *Mathematical Education for Teaching,* 3, 2, 36-40.

SHUARD, H.B. (1979): Language and reading in Mathematics. Unpublished paper BSPLM Meeting, Keele, January.

SKEMP, R.R. (1979): *Intelligence, Learning and Action.* Chichester: John Wiley.

STRAKER, A. (1978): 'The general adviser's role in primary mathematics. *Journal of NAIEA',* No. 9, Autumn, 12-13.

TRIVETT, J.V. (1977): 'Which researchers help teachers do their job?', *Mathematics Teaching,* No. 78, March, 39-43.

TROWN, E.A. and LEITH, G.O.M. (1975): 'Decision Rules for Teaching Strategies in Primary Schools: Personality - Treatment Interactions', *British Journal of Educational Psychology,* 45, 130-40.

WARD, M. (1979): *Mathematics and the 10-year-old,* (Schools Council Working Paper 61). London: Evans/Methuen Educational.

WEBB, N.L. (1980): The Core Curriculum, School Variables and Pupil Performance. Unpublished paper presented at the Annual Meeting of the American Educational Research Association, Boston, Mass., April.

WESTWOOD, L.J. (1967): 'The Role of the Teacher - I', *Educational Research,* 9, 122-34.

WILSON, B. (1962): 'The teacher's role - a sociological analysis', *British Journal of Sociology,* 13, 1, 15-32.

WITKIN, R.W. (1974): 'Social class influence on the amount and type of positive evaluation of school lessons' In: EGGLESTON, J. (Ed) *Contemporary Research in the Sociology of Education*. London: Methuen 302-24.

WRAGG, E.C. (1967a): 'The Lancaster study: its implications for teacher training', *British Journal of Teacher Education*, 2, 3, 281-90.

WRAGG, E.C. (Ed) (1976b): *Teaching Mixed Ability Groups*. London: David and Charles.

YATES, J. (1978): Four Mathematical Classrooms: An enquiry into teaching method. Faculty of Mathematical Studies, The University of Southampton.

YOUNG, M. (Ed) (1971): *Knowledge and Control*. London: Collier-Macmillan.

Appendix
Original Recommendations made to the Cockcroft Committee

In this chapter we present our recommendations based on our analyses of the research surveyed. For clarity and convenience the recommendations are grouped under the following headings:
(1)   Initial training
(2)   In-service training
(3)   Resource allocation
(4)   Research needs

1.   INITIAL TRAINING

1.A.   *TEACHERS' PERCEPTIONS OF MATHEMATICS*

1.A.1.   Student teachers should be given a broader perspective for their mathematics work, and more stress should be laid on the social and human context of mathematical knowledge.

1.A.2.   Their initial training should aim at increasing their confidence in mathematics through emphasizing the social and creative dimensions of the subject.

1.A.3.   They should be given, particularly, opportunities to engage more in discussions about their mathematics and to experience the negotiation of meaning, consciously.

1.A.4.   In their teaching practice they should be encouraged to develop lessons which allow for discussion, and they should be made aware of the importance of appropriate modelling behaviour.

1.B.   *CURRICULAR DECISION-MAKING OF PRIMARY TEACHERS*

1.B.1.   The initial training of primary teachers should lay greater stress on the curriculum decision-making aspect of their work.

1.B.2.   This should involve not just preparation to take such decisions but also information and understanding about the consequences of such decisions.

1.B.3.   Values-clarification exercises are necessary if teachers are to be aware of how their own values affect their decisions and their teaching methodology.

1.C.   THE INDIVIDUAL LEARNER AND MATHEMATICS

1.C.1.   Student teachers should be made aware of those superficial factors upon which judgments of pupils'
abilities are often made and of the effects such
judgments have.

1.C.2.   They should be encouraged to develop a 'researcher'
stance with regard to their pupils so that they will
decrease the likelihood of their making superficial
judgments.

1.C.3.   Primary student teachers should be particularly
encouraged to look for mathematical potential in their
pupils, and to develop ways in which they might
realize that potential.

1.C.4.   Secondary student teachers need particular help in
working with less able pupils, and in identifying
and encouraging the strengths and abilities they do
have.

1.C.5.   Student teachers generally should be encouraged to
develop a clear rationale for 'individualization'
which will help them to avoid making ad hoc decisions
about individual learners.

1.D.   CONSTRAINTS UPON PUPILS LEARNING MATHEMATICS

1.D.1.   Student teachers need to be made aware of the
'visibility' of mathematics and the ease with which
success and failure can be seen in the classroom.

1.D.2.   They need to be made aware moreover of the consequences of this visibility, particularly in the
area of pupils' attitude development.

1.D.3.   They should pay particular attention to this aspect
with respect to girls, and less able pupils, who
seem particularly discriminated against by this
visibility.

1.D.4.   They should be encouraged to develop their own
methods for reducing visibility, by perhaps
emphasizing more individual work with pupils and by
discouraging competitive aspects of tasks.

1.E.   TEACHER STRESS

1.E.1.   Student teachers should be made more aware of what
they can expect to happen in a classroom.

1.E.2.   There is a need to identify and discuss factors that
can often lead to teacher stress.

1.E.3.   Particular attention should be paid to *teacher*
behaviours which can lead to stressful situations,
such as those relating to 'visibility' above, or those
resulting from a superficial judgment of a pupil.

1.E.4.   Awareness of their own modelling behaviours needs to
be encouraged and also the avoidance of 'credibility'
crises.

1.E.5.   In all their initial training, student teachers
          should experience both real teaching situations and
          simulated situations. In simulations, role-play, and
          similar activities, many aspects of the social dimen-
          sion can be explored and experienced. The importance
          of many of the previous recommendations can be
          conveyed better through role-play, for example,
          than through discussion, though discussion of the
          role-play itself can enable the student teachers
          to talk through their worries, their values and
          their rationales.

2.   IN-SERVICE EDUCATION

2.A.   *ROLE OF THE PRIMARY HEAD TEACHER AND MATHEMATICS
          CO-ORDINATOR*

2.A.1.   Primary head teachers appear to need more guidance
          in how to take a more positive role in curricular
          leadership. This may entail study of mathematics
          curriculum theory as well as of leadership and
          support roles, the need for regular staff meetings
          and similar ideas.
2.A.2.   Mathematics co-ordinators in primary schools should
          be given special priority in in-service education,
          with similar guidance to that suggested above.
2.A.3.   As one area of difficulty is likely to be a conflict
          between the head and the co-ordinator in the school,
          opportunities should be sought to involve them in
          joint in-service programmes.

2.B.   *ROLE OF THE HEAD OF DEPARTMENT IN SECONDARY SCHOOLS*

2.B.1.   Much of the quality of secondary mathematics educa-
          tion rests with the heads of department, and they
          must be given priority in in-service education.
2.B.2.   Courses should be offered for those wishing to apply
          for head of department posts as well as for those
          already holding them.
2.B.3.   Emphasis in such courses should move from broadening
          their mathematical understanding to aspects of
          leadership, support, management, and within-
          institution negotiation.
2.B.4.   The importance of the department as a unit cannot be
          stressed enough in such in-service work and there is
          a good case for making much of this work school-based.

2.C.   *THE PROBATIONARY YEAR OF SECONDARY MATHEMATICS
          TEACHERS*

2.C.1.   A greater emphasis in in-service education should be
          given to the probationers than is apparent at present.
2.C.2.   There would appear to be a good case for making such
          in-service school-based and, essentially, the

81

responsibility of the head of the mathematics department.

2.C.3. Whatever organisation is used, there appears to be a great need to co-ordinate the activities of the various people concerned with probationers.

2.D. *LINKS WITH OTHER SUBJECTS*

2.D.1. It does appear that mathematics departments do not link well with other subject departments and this aspect needs more exploration.

2.D.2. It is not necessary for this to be considered part of the head of department's role, but it could well be taken on by another member of the department.

2.D.3. Joint in-service education with other departments and other advisers would seem to be necessary also, to consider curricular matters, materials used and other teaching matters.

2.D.4. In primary schools, the mathematics co-ordinators should be responsible for advising about such links, and this aspect should form part of their in-service training.

2.E. *OTHER PERSONNEL*

2.E.1. As has been indicated, many other school personnel influence mathematics teaching and every opportunity should be taken to ensure awareness of this fact.

2.E.2. Particular emphasis needs to be given to this in the in-service education of heads, deputy heads, time-table planners, resource personnel and teachers of other subjects (particularly user subjects).

3. RESOURCE ALLOCATION

3.A. *PRIMARY SCHOOL GUIDANCE*

3.A.1. There is in our view a great need for the provision of more mathematics co-ordinators in primary schools. This is a priority because of the apparent shortage of advisory staff and the inattention being given to the problem by head teachers.

3.A.2. Such co-ordinators should be reasonably experienced teachers who have received special training in mathematics curriculum areas.

3.A.3. Additionally there should be more advisors provided with speical responsibility for mathematics at primary level.

3.B. *MATHEMATICS SPECIALIST AREAS*

3.B.1. At secondary school the main lack, apart from the shortage of well qualified teachers, is in the provision of mathematics specialist areas.

3.B.2. Such areas should include at least linked classrooms and a resource centre.

3.B.3. Additionally, in new schools, it would be sensible to locate a computer room near the area, and to have a variety of teaching accommodation rather than several identical classrooms.

3.C.   *MATERIALS*

3.C.1. Reference has been made to the weaknesses in teaching the less able at secondary school, and there does exist a need for more relevant materials for that age and ability level.

3.C.2. There should be more sharing of resources between the mathematics and remedial departments, to the mutual benefit of both.

3.C.3. In primary schools there may well be a lack of suitable materials for extending the more able, and where it exists, this deficiency should be remedied.

3.C.4. Achievement test provision would appear to be reasonably satisfactory, but ways of assessing pupils to determine mathematical *potential* and *aptitude* need to be developed for use in primary school.

3.C.5. Micro-computer provision needs to be increased to ensure that all pupils leave school having experienced some computer work.

4.   RESEARCH NEEDS

4.A.   *LEADERSHIP*

4.A.1. There is a need for more knowledge about the roles and functions of heads of department in secondary school.

4.A.2. A start could be made by analysing 'Ten Good Departments' and perhaps also 'Ten Bad Departments'.

4.A.3. More in-school research needs to be undertaken concerning resource allocation decisions, time-tabling decisions, in-school negotiation, etc.

4.A.4. Primary school curricular leadership also needs much more analysis with a focus on the relationships between the roles of heads/co-ordinators/advisers/teachers.

4.A.5. In-service education could be aided greatly by studies of teachers' curricular decision-making habits.

4.B.   *PROBATIONARY YEAR*

4.B.1. The particular plight of the mathematics probationer needs more clarification.

4.B.2. More knowledge of their duties and perceived pressures would be beneficial, together with

information about the advice and help they felt
they received.

4.B.3.   A useful project would be to study the detailed
case histories of probationers who have had their
period of probation extended.

4.C.   *CONSTRAINTS UPON PUPILS*

4.C.1.   We need more knowledge about the particular effects
on the pupils of the visibility of the criteria for
judging mathematics learning.

4.C.2.   It would be particularly useful to explore the
relationship between the teacher's methodology and
those effects.

4.C.3.   Generally more understanding of 'individualization'
would be valuable, whether from the teachers'
perspective, e.g. what rationales for individualization they adopt, or from the pupils' perspective,
e.g. which methods and approaches favour particular
individuals.

4.C.4.   The within-classroom constraints on girls' mathematical learning needs greater clarification. Of
particular interest would seem to be the 'fear of
success' construct devised in the U.S.A. to help
explain girls' reluctance to perform well in
mathematics classes.

4.D.   *TEACHER PLANNING*

4.D.1.   One promising research area which should be studied
in this country concerns teacher planning, and the
role that planning plays in controlling much of what
happens in classrooms.

4.D.2.   One particular hypothesis which could be explored is
that planning leads to an insensitivity to individual
pupils and a lack of flexibility in responding to
them.

4.D.3.   An analysis of teacher planning activities and how
these vary at both primary and secondary levels could
be most instructive.

4.E.   *IDENTIFICATION OF MATHEMATICAL POTENTIAL*

4.E.1.   There appears to exist a great need for a study of
teacher judgments of pupils' mathematical potential.

4.E.2.   It would also be of interest to compare the bases for
these judgments, and their validity, between primary
and secondary teachers.

4.E.3.   Further information is required about the use by
teachers of published, and school-produced, tests,
and the extent to which they feel that the tests
identify mathematical potential, as opposed to
mathematical achievement.